Get Ready for Biology

Get Ready for Biology

Adapted from *Get Ready for A&P*,

Lori K. Garrett, Danville Area Community College

San Francisco Boston New York Cape Town Hong Kong London Madrid
Mexico City Montreal Munich Paris Singapore Sydney Tokyo Toronto

Editor-in-Chief: Beth Wilbur
Director of Development: Deborah Gale
Developmental Editor: Suzanne Olivier
Associate Editor: Leata Holloway
Executive Marketing Manager: Lauren Harp
Production Supervisor: Lori Newman
Production Service: Martha Emry
Cartoonist: Kevin Opstedal
Cover and Text Designer: Seventeenth Street Studios
Photo Researcher: Travis Amos
Compositor: Seventeenth Street Studios
Proofreader: Nicole Wisner
Indexer: Sallie Steele
Manufacturing Buyer: Stacy Jenson
Cover and Text Printer:
RR Donnelley and Sons

www.aw-bc.com

ISBN-10: 0-321-50057-1
ISBN-13: 978-03215-0057-1

Contents

Preface

Welcome to the fascinating world of biology, the study of life! Most students take biology because they have to—it's a required part of their educational curriculum. But we hope you quickly discover how amazing life on Earth is and become intent on learning all that you possibly can about it. After all, you are *part* of life.

For many reasons, students sometimes do not succeed in their first biology class. If you're reading this Preface, you probably have a strong desire to succeed, and you know the competition for admission into numerous educational programs is increasing. You are likely keenly aware that you can't just pass your classes—you need to do quality work and truly master the course content. This book offers you the opportunity to enhance your performance in this rigorous course. It is designed to help you gear up to be successful.

You may be using this book before your course officially begins. Your instructor may have assigned it to you as homework during the first week or two of class. Perhaps you are using it on your own and will come back to it periodically throughout the semester as you move along in your course. However you use it, the purpose of the book remains the same. The goal is to help you get a strong start in biology, and to master the material not just for exams, but for your future as well. *Get Ready for Biology* contains six interactive chapters that engage you every step of the way. You'll read, but you'll also frequently do activities.

The book starts with basic study skills in **Chapter 1**. As with any course, you will get out of biology what you put into it, and it will take significant time and effort on your part to succeed. This chapter helps you focus and manage your time so you can first find time to study, then use your study time effectively. After exploring different learning styles you can assess which style best fits you, then discover specific study strategies that complement your preferred style. You'll assess your current habits as a student and learn specific tips and strategies to help you study better. Specific tips will help you write your notes, read textbooks, and take tests.

Chapter 2 covers basic math skills. Biology is a science, and all science involves at least some math. This chapter takes you from basic math operations through reading and interpreting numerical information in graphs and tables—the math you'll need for a head start in your course.

Many of the words in your biology class will sound foreign to you, and well they should! Most of the terms come from Latin or Greek. Knowing the terms underlies all aspects of learning in this class. In **Chapter 3**, Terminology, we look at how the words are built and learn some simple tricks that will rapidly expand your biology vocabulary.

The second half of the book gets more specific and parallels some of what you will cover in the first few chapters of your biology textbook. In **Chapter 4**, we cover biology basics: the scope of biology and some general biological principles that help explain how life works. We'll also take a quick look at how scientific inquiry is done.

In **Chapter 5**, we tackle chemistry. This chapter gives you the basics, from atoms to organic molecules. We discuss some neat tricks for gaining information from the Periodic Table, and see how atoms join together to form molecules. If you can understand bumper cars, you can understand bonding!

Finally, in **Chapter 6**, we explore cells. Every living thing is made up of one or more cells, and almost everything that happens in living organisms occurs inside a cell. We discuss basic cell structure, the cell life cycle, and cell reproduction.

Now that you know the roadmap for this book, let's explore the stops you'll find along the way. Here are the special features in each chapter, designed to keep you involved and to make you a better student in biology:

- ***Your Starting Point*** tests your grasp of the chapter content before you start. Answers are provided for all of these except in Chapter 1, where the answers are personal.
- ***Quick Check*** asks you to recall or apply what you just read, to keep your eyes from scanning the page while your brain is on vacation. The answer is provided on the same page.
- ***Picture This*** asks you to visualize scenarios and then answer questions about them to help you better understand the topics.
- ***Time to Try*** is a simple experiment or quick assessment in which you perform an active exercise.
- ***Why Should I Care?*** highlights the relevance of the material so you understand its importance in the big picture.
- ***Reality Check*** assesses whether you really "got" the material.
- ***Keys*** highlight main themes or statements for reinforcement and easy review.
- ***Running Words*** list key terms from the chapter to help you start your own running vocabulary list by writing each term in a notebook, then defining it.
- ***What Did You Learn?*** end-of-chapter quizzes may include short answer, multiple choice, or matching exercises. The answers appear at the end of the book.

Now it is time to dig in. So get comfortable, and *Get Ready for Biology!*

Acknowledgments

Get Ready for Biology is adapted from *Get Ready for A&P*, a book written by Lori Garrett of Danville Area Community College. Her simple, straightforward explanations and suggestions for ways to help students succeed are evident throughout the adaptation, and we thank her for her talents and her compassion for students.

Also deserving thanks are those who helped make *Get Ready for A&P* happen: the Benjamin Cummings editorial team, Jessica Brunner, Claire Alexander, and Serina Beauparlant; the production team, and in particular Wendy Earl and Travis Amos; designer Randall Goodall and his composition team at Seventeenth Street Studios; and cartoonist Kevin Opstedal.

Thanks also go to the team that worked on the adaptation, *Get Ready for Biology*, to publish it in record time: developmental editor Suzanne Olivier, who adapted the anatomy and physiology version for biology students; associate editor Leata Holloway, who teamed up with production supervisor Lori Newman; and Randall Goodall and Seventeenth Street Studios.

Finally, we owe a great debt of gratitude to all of our reviewers, who are listed on the following page. They provided thoughtful and candid feedback, which helped make the book stronger and more useful to students.

REVIEWERS

Erin Amerman
Santa Fe Community College

Vince Austin
Bluegrass Community & Technical College

Claudie Biggers
Amarillo College

Margaret Creech
Laramie County Community College

Terry Harrison
Arapahoe Community College

Clare Hays
Metropolitan State University

Maurice Heller
El Paso Community College

Mark Hubley
Prince George's Community College

Catherine Hurlbut
Florida Community College, Jacksonville

Jody Johnson
Arapahoe Community College

Johanna Kruckeberg
Kirkwood Community College

Ken Malachowsky
Florence-Darlington Technical College

Elaine Marieb
Holyoke Community College

Judy Megaw
Indian River Community College

Claire Miller
Community College of Denver

Amy Nunnally
Front Range Community College

Wayne Seifert
Brookhaven College

Joan C. Sharp
Simon Fraser University

Alan Spindler
Brevard Community College

Dieterich Steinmetz
Portland Community College

Yong Tang
Front Range Community College

Marty Tayler
Cornell University

Deborah Temperly
Delta College

Jennifer van de Kamp
Front Range Community College

1 Study Skills

Training Your Brain

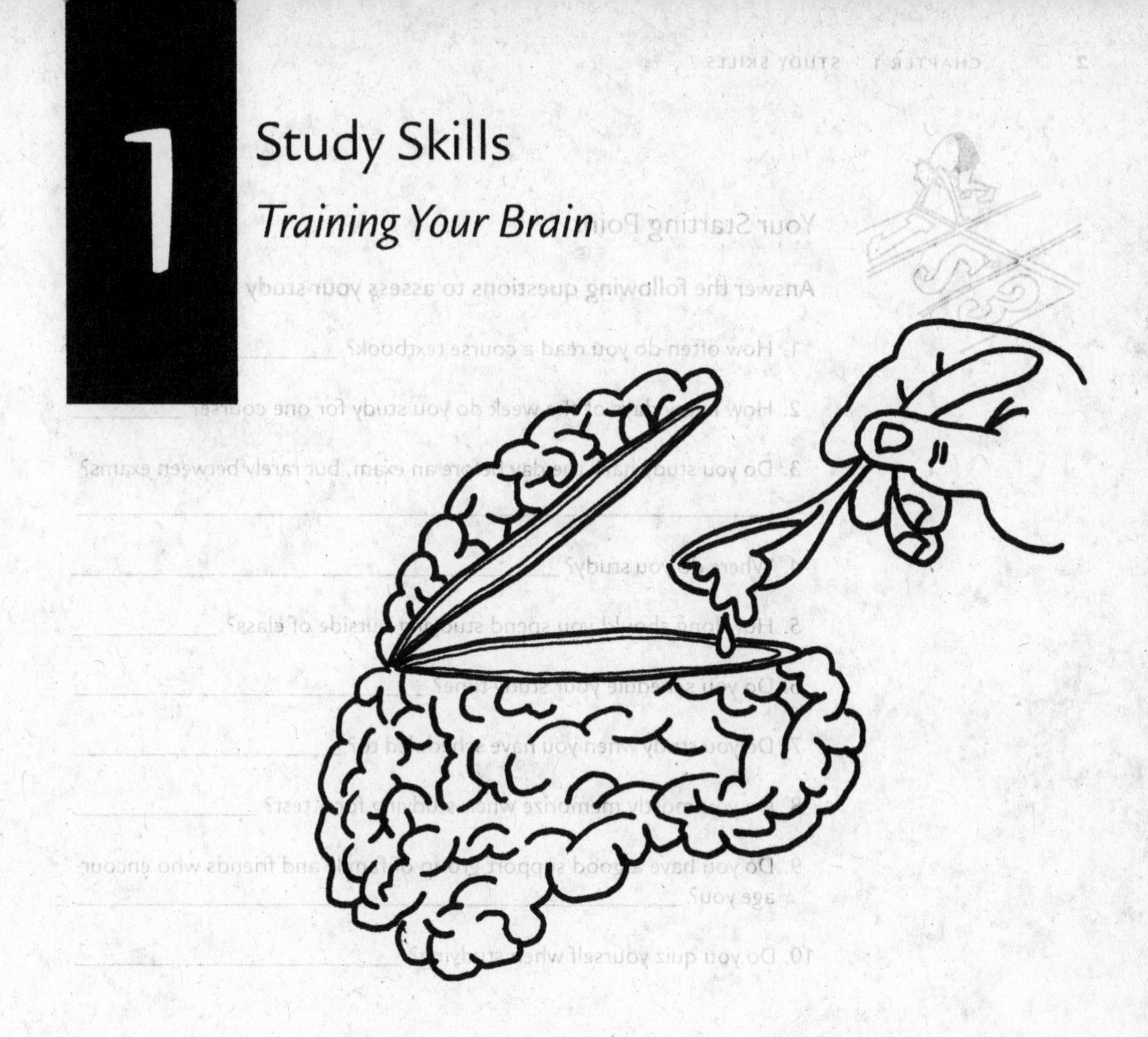

When you complete this chapter, you should be able to:

- Understand your preferred learning style and study strategies that emphasize it.
- Have skills that will help you get the most benefit from lectures, labs, and readings.
- Have a written schedule that includes adequate study time.
- Know how to prepare well for an exam.
- Understand that you are ultimately accountable for your own success or failure.

Your Starting Point

Answer the following questions to assess your study habits.

1. How often do you read a course textbook? ____________
2. How many days of the week do you study for one course? ________
3. Do you study hard the day before an exam, but rarely between exams? ____________
4. Where do you study? ____________
5. How long should you spend studying outside of class? ____________
6. Do you schedule your study time? ____________
7. Do you study when you have scheduled to? ____________
8. Do you mostly memorize when studying for a test? ____________
9. Do you have a good support group of family and friends who encourage you? ____________
10. Do you quiz yourself when studying? ____________

Welcome to the exciting and sometimes challenging world of biology, the study of life. You will quickly discover how amazing the living world is—a curious marvel of complexity that we hope you will be fascinated by. Interest in your subject matter always makes it much easier to learn.

Still, no matter how exciting your explorations may be, your course may, at times, seem rigorous and demanding. You've taken a great first step by turning to this book to jump-start your studies. This book is meant to help you enter the course with a well-planned strategy for success and with confidence in your basic science knowledge. The purpose of this chapter is to help you "train your brain" to make your learning process easier and more efficient.

Answers: Answers will be individualized, except for #5—you should spend 2–3 hours studying for each hour of class time.

Why Should I Study Biology?

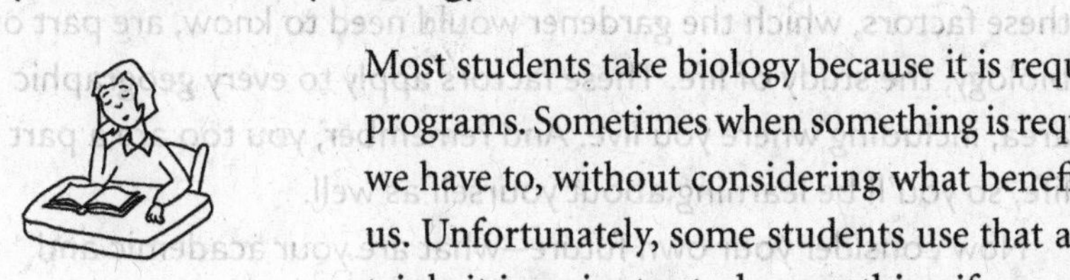

Most students take biology because it is required for their educational programs. Sometimes when something is required we do it only because we have to, without considering what benefits the task might hold for us. Unfortunately, some students use that approach for biology. Certainly it is easier to study something if you understand why it matters, and this course is no exception.

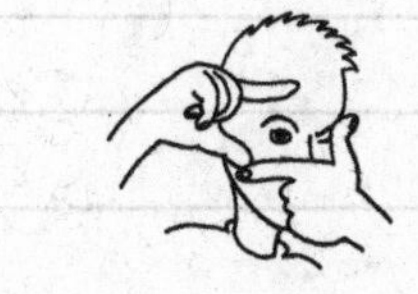

PICTURE THIS

Suppose that your parents have a back yard. Their yard has flower beds and a small vegetable garden. But something has been eating the flowers and the beans, and your parents can't figure out what animal is responsible. Caterpillars? Deer? Gophers? They've tried everything they can think of. They asked you and others for ideas, but no one can figure it out. Finally, they decide to bring in a professional gardener to help them. What knowledge will the gardener need to identify and stop the animals from eating your parents' garden?

____

____

____

Why does the gardener need to fully understand biology?__________

____

____

To understand what the plants need in order to grow, what animals may be living in the area and like to eat those plants, and what would deter the animals and save the plants, the gardener will need to understand not only plants and animals but also how they live together in the environment, forming a community. How does the amount of rain and sun affect the plants and animals? What nutrients do the plants need? What animals might be in competition to eat the plants? How does the season of the year affect the animals and plants? Do the animals have young they need to feed? What impact do the actions of humans have on this community? All of these factors, which the gardener would need to know, are part of biology, the study of life. These factors apply to every geographic area, including where you live. And remember, you too are a part of life, so you'll be learning about yourself as well.

Now consider your own future—what are your academic and career goals?

____

____

Why will you need to know biology? ______________________

____

To Thine Own Self Be True: **Learning Styles**

What *is* the best way to learn these subjects? A tremendous amount of research has explored how people learn, and there are many opinions. One common approach considers which of the senses a learner relies on the most—sight, sound, or touch:

- Visual learners learn best by *seeing.*
- Auditory learners learn best by *hearing.*
- Tactile (kinesthetic) learners learn best by *doing.*

TIME TO TRY

Let's uncover your learning style.

1. Look at **Table 1.1**. Read an activity in the first column, then read each of the three responses to the right of that activity.
2. Mark the response that seems most characteristic of you.
3. After doing this for each row, you are ready for your totals. Simply add all the marks in each column and write the total in the corresponding space in the bottom row.

4. Next look at your numbers. You will likely have a higher total in one column. That is your primary learning style. The second highest number is your secondary learning style.

My primary learning style is: ______________________

My secondary learning style is: ______________________

TABLE 1.1 Assessing your learning style.

Activity	Column 1	Column 2	Column 3
1. While I try to **concentrate** . . . 	I grow distracted by clutter or movement, and I notice things in my visual field that other people don't.	I get distracted by sounds, and I prefer to control the amount and type of noise around me.	I become distracted by commotion, and I tend to retreat inside myself.
2. While I am **visualizing** . . . 	I see vivid, detailed pictures in my thoughts.	I think in voices and sounds.	I see images in my thoughts that involve movement.
3. When I **talk to someone** . . . 	I dislike listening for very long.	I enjoy listening, or I may get impatient to talk.	I gesture and use expressive movements.
4. When I **contact people** . . . 	I prefer face-to-face meetings.	I prefer speaking by telephone for intense conversations.	I prefer to interact while walking or participating in some activity.

▶

TABLE 1.1 Assessing your learning style, continued.

Activity	Column 1	Column 2	Column 3
5. When I **see an acquaintance** . . .	I tend to forget names but usually remember faces, and I can usually remember where we met.	I tend to remember people's names and can usually remember what we discussed.	I tend to remember what we did together and may almost "feel" our time together.
6. When I am **relaxing** . . .	I prefer to watch TV, see a play, or go to a movie.	I prefer to listen to the radio, play music, read, or talk with a friend.	I prefer to play sports, make crafts, or build something with my hands.
7. While I am **reading** . . .	I like descriptive scenes and may pause to imagine the action.	I enjoy the dialogue most and can "hear" the characters talking.	I prefer action stories, but I rarely read for pleasure.
8. When I am **spelling** . . .	I try to see the word in my mind or imagine what it would look like on paper.	I sound out the word, sometimes aloud, and tend to recall rules about letter order.	I get a feel for the word by writing it out or pretending to type it.
9. When I **do something new** . . .	I seek out demonstrations, pictures or diagrams.	I like verbal and written instructions, and talking it over with someone else.	I prefer to jump right in to try it, and I will keep trying and try different ways.

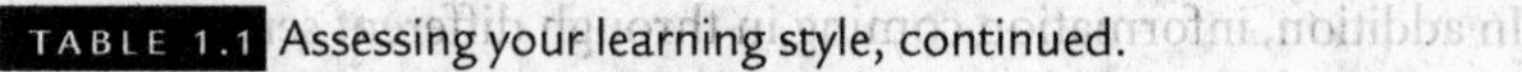
TABLE 1.1 Assessing your learning style, continued.

Activity	Column 1	Column 2	Column 3
10. When I **assemble something** . . . 	I look at the picture first and then, maybe, read the directions.	I like to read the directions, or I talk aloud as I work.	I usually ignore the directions and figure it out as I go along.
11. When I am **interpreting someone's mood** . . . 	I mostly look at his or her facial expressions.	I listen to the tone of the voice.	I watch body language.
12. When I **teach others how to do something** . . .  	I prefer to show them how to do it.	I prefer to tell them or write out how to do it.	I demonstrate how it is done and ask them to try.
TOTAL:	**Visual:**	**Auditory:**	**Tactile/Kinesthetic:**

[Source: Courtesy of Marcia L. Conner, www.agelesslearner.com]

Now that you know your primary and secondary learning styles, you can design your study approach accordingly, emphasizing activities that use your preferred senses. Look closely at your scores, though. If two scores are rather close, you already use two learning styles well and will benefit from using both of them when studying. If your high score is much higher than your other scores, you have a strong preference and should particularly emphasize that style. Most people use a combination of learning styles.

In addition, information coming in through different senses reaches different parts of your brain. The more of your brain that is engaged in the learning process, the more effective your learning will be, so try strategies for all three styles and merely emphasize your preferred style over the others. You'll know which strategies work best for you. We'll consider some strategies that you might try for each style; these ideas are summarized for you in **Table 1.2.**

TABLE 1.2 The three learning styles and helpful techniques to use in your studies.

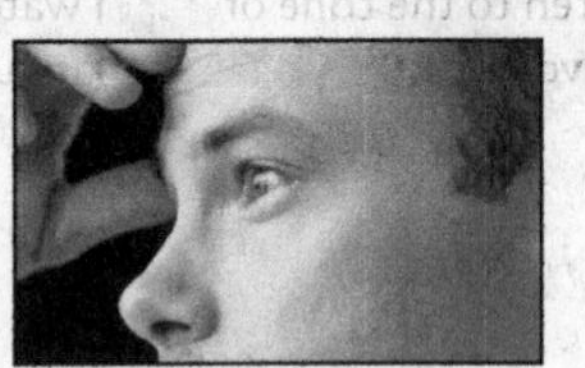
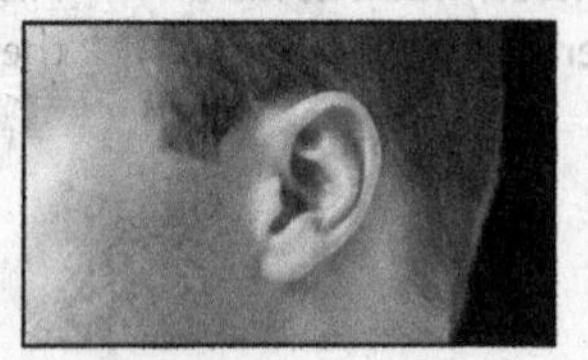
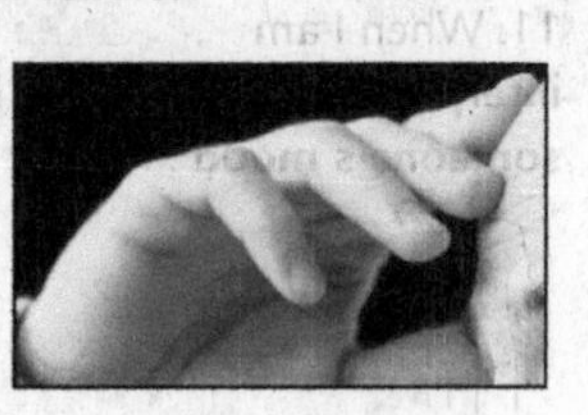

	Visual	Auditory	Tactile
Techniques to use	❑ Sit close to the teacher. ❑ Take detailed notes. ❑ Draw pictures. ❑ Make flow charts. ❑ Use flash cards. ❑ Focus on the figures, tables, and their captions. ❑ Try coloring books and picture atlases. ❑ Use visualization.	❑ Listen carefully to your teacher's voice. ❑ Read the textbook and your notes out loud. ❑ Tape record lectures and listen to them later. ❑ Listen during class instead of writing notes. ❑ Work in a study group. ❑ Discuss the material with others.	❑ Highlight important information while reading. ❑ Write your own notes in class and while reading the textbook. ❑ Transfer your notes to another tablet or type into your computer. ❑ Doodle and draw as you read. ❑ Build models of biological structures. ❑ Create and conduct your own experiments. ❑ Hold your book while reading. ❑ Walk or stand while reading. ❑ Use biology coloring books. ❑ Use flash cards.

VISUAL LEARNERS

If you are a **visual learner**, you rely heavily on visual cues. You notice your teacher's mannerisms, expressions, gestures, and body language. Seeing these cues is especially helpful, so sit at the front of the classroom, close to the teacher. You tend to think in pictures and learn well from visual aids such as diagrams, illustrations, tables, videos, and hand-outs. Here are some strategies for you:

- In class, take detailed notes and make sketches.
- When studying on your own, draw pictures that relate to the information, make flow charts and concept maps, use flash cards, focus on illustrations and tables in your textbook, and read the captions that accompany them.
- Use biology coloring books and picture atlases.
- Use mental visualization of the material you are studying and imagine yourself acting out processes. For example, to learn the carbon cycle, you might imagine yourself as a carbon atom going through the cycle.

AUDITORY LEARNERS

If you are an **auditory learner**, you learn well from traditional lectures and discussion. You listen carefully to your teacher's vocal pitch, tone, speed, and mannerisms. Material that you struggle with while reading becomes clearer when you hear it. Here are some strategies for you:

- Read the textbook and your notes out loud.
- Tape record the lectures so you can listen to them later. Taping lectures also allows you to listen during class instead of focusing on writing, which is less beneficial for you.
- Work in a study group, and discuss material with your teacher, lab group, and friends.

TACTILE LEARNERS

If you are a **tactile learner**, you learn best by actively participating and doing hands-on activities. You may become bored easily in class from sitting still too long and start fidgeting or doodling. You need to do something physical while studying and learning. Here are some strategies for you:

- Try using a marker to highlight important information while you are reading.
- Write out your own notes in class and while reading the textbook. Later, transfer your notes to another tablet or type them into your computer.
- Draw pictures of appropriate material as you read.
- Build models of structures using clay or other materials.
- Create and conduct your own experiments.
- Hold your book and walk while reading.
- Use biology coloring books.
- Make and use your own flash cards.
- Keep your hands and your mind busy together.

WHY SHOULD I CARE?

Understanding your own learning style allows you to develop more effective and efficient study techniques that take advantage of your sensory preferences. By emphasizing your preferred learning style, the material will be easier to learn and will stay with you longer.

✔ **QUICK CHECK**

Homemade flash cards would be most beneficial to which two learning styles? ______________ and ______________.

How could they be used to benefit a learner of the third style?

____

Answer: They would benefit visual and tactile learners. Reading them out loud would benefit auditory learners.

Putting Your Best Foot Forward: **Getting Ready**

Many students mistakenly wait until the first lecture to start thinking about class. The key to starting your semester well is to be organized and ready when you enter the classroom. This takes advance planning, but the time invested will save you tremendous time while the semester is underway.

PUTTING IT IN WRITING

As the semester begins—if not before—you should get organized, and that begins with making a commitment to yourself. Too often we begin a project without setting goals in advance. If you set a goal, you enter with a purpose and a direction. If you do not set a goal, it's too easy to just go along and see where you end. Take time to think about your goals for the semester. They should be both specific and attainable. Be realistic. For example, it may not be realistic to set the goal of always having the highest score in class, but a goal of getting an A in the class might be attainable. Once you've decided on your goals, write them down to give them more importance. Once you've written them, be firmly committed to them. To reinforce these goals, write them on an index card and place it in a prominent location in your study area so you'll see them every day.

TIME TO TRY

Set three main goals for yourself in this class, and write them below. Explain why achieving each goal is important to you.

Goal 1: ______________________________

It is important to me because: ______________________________

____

Goal 2: ______________________________

It is important to me because: ______________________________

____

Goal 3: ______________________________

It is important to me because: ______________________________

____

PULLING IT ALL TOGETHER

Have you ever seen a fellow student show up for a test with no pen or pencil? Don't let that happen to you. The more organized you are, the more efficient you will be, so let's organize what you will need for class. Categorize the items by what you take to class every day, what remains at home in your study spot, and optional items that are nice, but nonessential, additions. Use the checklist provided for you in **Table 1.3**. Search your house and you'll likely find that you have many of these items already. Most items can be bought at your college bookstore, but many are available at discount stores. We will discuss some of these items specifically.

You'll be going back and forth to class a lot, so it is most efficient to keep all the items you might need for class in one place. To haul them, most students use a book bag, backpack, or briefcase. An advantage to using one of these is that you can load it up with the essentials so that they are always ready to walk out the door with you. Let's discuss some of the items to pack.

You need a pocket-sized day planner that has plenty of room for writing and that you can keep with you at all times. Or you may opt for a personal organizer portfolio or an electronic organizer. Select one you like,

TABLE 1.3 Organizer's checklist.

Item	✎✗
To take to class each day:	
Book bag/backpack/rolling carrier	
Textbooks/lab manuals	
Pocket-sized day planner	
To Do List	
Separate notebooks for each course	
Copy of class schedule with buildings and room numbers	
Several blue or black ink pens	
Several #2 pencils	
Small pencil sharpener	
2–3 colored highlighter pens	
Small stapler	
Grade record sheet for each course	
Calculator	
At home:	
Master calendar	
Separate file or folder for each course	
Loose notebook paper	
Index cards for making flash cards	
Computer paper (if I have a computer)	
More writing utensils (pens and pencils)	
Stapler	
Calculator	
Scissors	
Paper clips	
Optional:	
Personal organizer	
Biology coloring book	
Colored markers/pencils	
Small tape recorder to record lectures/readings	
Recording tapes	
Extra batteries	
Biological dictionary	

because you'll use it every day. In it, write all important dates you already know—when classes begin, holidays, last day to withdraw from a class, when finals begin. Enter all class times, your work schedule, and any other known time commitments. Try to keep your day planner current so you always know how your time is being spent and can plan ahead.

If it's not part of your day planner, you need a separate To Do List. You will write all assignments and due dates on this list. You want one single To Do List for all of your classes as well as non-school activities, because they must all be done from the same pool of time. Writing them down allows you to view the entire list and review the deadlines for each item so you can easily prioritize, doing the assignments in the order in which they are due.

Maintain a record of all grades you receive (**Figure 1.1**). For each graded item, list what it is, when you turned it in, when you got it back, how many points you received, how many points were possible, and any additional notes. Once you know how your grade will be determined for the course, you can use this to keep track as you go along. It also provides a backup in case there is any confusion later about your grade or your work.

Graded item	Date turned in	Date returned	My score	Possible points	Notes
Lab 1	9/6	9/13	10	10	Worked with Emily, Mike, Tom
Quiz 1	9/7	9/9	18	20	Study terms again
Lab 2	9/13	9/16	6	10	Messed up the math!
Pop Quiz	9/14	9/16	5	5	From yesterday's lecture. I was ready!
Quiz 2	9/21	9/25	19	20	Forgot to answer one question!

FIGURE 1.1 A sample grade record for keeping track of your progress.

Check with your instructor to see if you should bring your textbook to class. Typically, you may not need it in lecture, but you may in lab. If your course uses a lab manual, always take it to lab.

And for both lecture and lab, always carry the basics. You need a notebook for note-taking. If you come to class without a pen or pencil, it says you do not think that what the instructor is saying is important enough for notes. If you ask your instructor for a stapler before turning in an assignment, it says you threw it together with little thought. Pens, pencils, erasers, staples, paper, highlighters, colored pencils, index cards, paper clips—these are just some items you may find useful. Replenish your supply as needed.

Set up your home study space like a home office. Be sure to have all the essential office supplies on hand—plenty of writing utensils, paper, a stapler, and so on. A critical part of the home study area is the master calendar. There are large desktop versions and wall charts, for example. You could use a calendar feature on your computer, but the more visible the calendar is, the more often you will look at it. This calendar should be large enough to accommodate plenty of writing, so think BIG! Each day, you should add anything that you put in your day planner or on your To Do List to this master calendar. All time commitments should be entered, so also add all personal appointments and vacations. This is how you will schedule your life while in school, and the practice will likely stick with you far beyond that.

You may have already started your class before using this workbook. You certainly cannot do all these things before class at this point, but it is never too late to get organized. So, go get it together!

✔ **QUICK CHECK**

To be successful in class, your effort should start before class begins. What are some tasks you should do before the first day of class?

____

____

____

Answer: Set and write down your goals, organize the items you will need for class and at home, pack your carrier, start your day planner and master calendar, and organize your study space.

____

I Hate to **Lecture** on this, but Can You Hear Me Now?

Welcome to class! Imagine that it is the first day. You walk into class.

Where do you sit? ______________________________

Why do you sit there? ______________________________

The best seat in the house is front and center. Obviously not everyone can sit there, but you should arrive early enough to sit within the first few rows and as near to the middle as possible. You want an unobstructed view of the instructor and anything he or she might show, because biology is often a very visual course. People sitting on the sides or in the back often do not want to be called on, or they want to be in their own space. They are not very engaged in the class. Don't let that be you. To succeed, you need to focus all of your attention on your instructor, minimize distractions, and actively participate. Instructors tend to teach to the middle of the room (**Figure 1.2**). In fact, if your instructor is right-handed and uses equipment, such as an overhead projector, that is positioned on the right, the instructor's focus shifts to his or her right. You want to see your instructor and you want your instructor to see that you are present, actively listening, and engaged.

Some instructors provide lecture notes so you can sit back and really think about what is being said. Notes or not, you need to get all the information you can from each lecture. Remember your learning style and use techniques that enhance it. We will discuss note-taking momentarily, but consider tape recording the lectures. That way you miss nothing, and you can listen to the tape repeatedly, rewinding as needed. Another good technique is to write out your own notes while listening to the tape, then listen again while reading your notes and making necessary corrections. This combination strongly reinforces the material.

Always try to preview the material that will be covered before going to class. This is as simple as lightly reading the corresponding sections in the textbook. You may not understand all that you read, but it will sound familiar and be easier to comprehend as your instructor covers it in class. This preview also helps you identify new vocabulary words.

While your instructor is lecturing, don't hesitate to raise your hand to ask a question or get clarification. Many students are shy and reluctant to speak in class—you may be doing them a favor! Avoid discussing

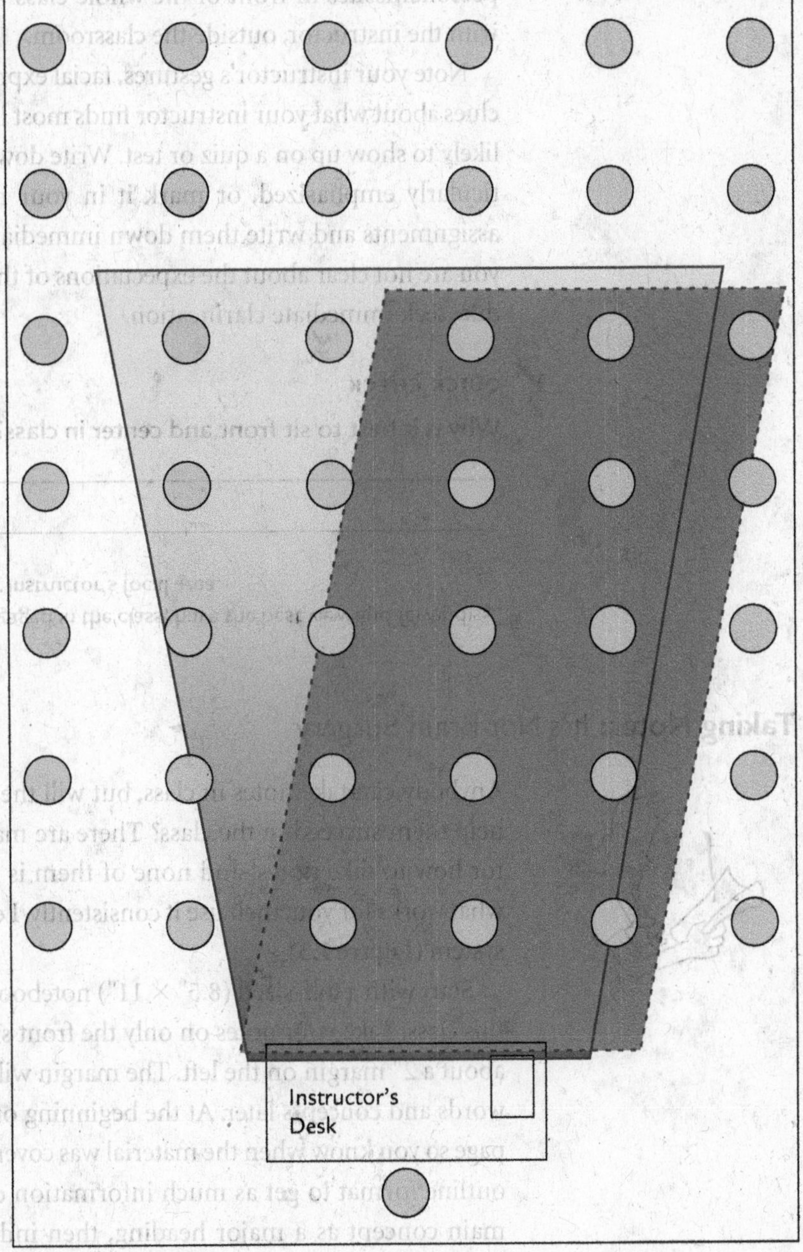

Instructor's Desk

FIGURE 1.2 Main areas of focus for an instructor. The larger shaded area shows where the instructor looks most often. The smaller shaded area shows how that focal area shifts to the right as the instructor uses equipment positioned to the right.

personal issues in front of the whole class—that is better done alone with the instructor, outside the classroom.

Note your instructor's gestures, facial expressions, and voice tone for clues about what your instructor finds most important. That material is likely to show up on a quiz or test. Write down any material that is particularly emphasized, or mark it in your notes. Listen carefully for assignments and write them down immediately on your To Do List. If you are not clear about the expectations of the assignment or when it is due, seek immediate clarification.

✔ **QUICK CHECK**

Why is it best to sit front and center in class?

____

____

Answer: You will be more engaged in the class, have the best view and fewer distractions, and be within your instructor's focal area.

Taking Notes: It's Not Brain Surgery

Anybody can take notes in class, but will the notes be good enough to help them succeed in the class? There are many strategies and models for how to take notes, and none of them is necessarily the best. Find what works for you, then use it consistently. Let's review one easy-to-use system (**Figure 1.3**).

Start with a full-sized (8.5" × 11") notebook that you will use just for this class. Take your notes on only the front side of the paper and leave about a 2" margin on the left. The margin will be used for marking key words and concepts later. At the beginning of class, date the top of the page so you know when the material was covered. During lecture, use an outline format to get as much information down as you can. Use the main concept as a major heading, then indent the information discussed on that topic. When that section ends, either draw a horizontal line to mark its end or leave a couple of blank lines. Don't try to write every word—just the main ideas—and put them in your own words. Instead of writing out every example, give a brief summary or a one- or

RECALL:	03/14/06
Note-taking	I. Note-taking tips
	A. Use outline format
	B. Be concise
	C. Get main ideas
Reviewing	II. Reviewing notes
	A. Review after class
	1. Fill in gaps
	2. Clean up
	3. Replay lecture in my mind
	4. Review within 24 hours—fresh in mind.
3 learning styles	III. Learning styles
	A. Visual—reread and add drawings
	B. Tactile—rewrite or type
	C. Auditory—read out loud or tape record

FIGURE 1.3 Sample of lecture notes using the outline style and leaving room in the left margin.

two-word reminder. Use abbreviations when possible, and develop your own shorthand. You can often drop most of the vowels in a word and still be able to sound it out later when reading it. Write legibly or your efforts will be useless later. Underline new or stressed terms and place a star or an arrow by anything that is emphasized. Be as thorough as you can, but you will need to write very quickly. The instructor will not wait for you to catch up, so speed is essential.

As soon as possible after class, read your notes and improve them as necessary. Add anything that's missing. Make them clearer and cleaner. Put the concepts in your own words. Next, use the left margin to summarize each section—the main concept, subtopics, and key terms. The latter column will be your "Recall" column. Once you are sure all of the key ideas are in the left column, you can cover the right side of the page—the meat of your notes—and quiz yourself on the main points listed in the left column. It makes an easy way to review.

But you are not finished—if you are a tactile learner, rewrite your notes in another notebook or type them into your computer. Visual learners might type and reorganize the notes. Auditory learners can read the notes out loud or record them. You can make flash cards from the key points and terms by writing the term on one side of an index card and its definition or use on the other side. You can add drawings. Review your notes as much as you can during the next 24 hours, while the lecture is still fresh in your mind.

TIME TO TRY

Look at the sample notes in Figure 1.3. Now practice: Take notes using this style while listening to a 1-hour TV show. Capture the conversations and action in words. You can't get every word down, so paraphrase—put it in your own words so the meaning still comes across. When you're finished, assess how you did.

Can you tell who was talking? ______________________

Do your notes make sense? ______________________

Did you capture the main ideas? ______________________

Did you keep up or fall behind? ______________________

Do you have breaks in your notes to separate the main conversations and action? ______________________

What can you do better when taking notes in class?

____

✔ **QUICK CHECK**

What should you do with your notes after class? ______________________

____

____

Answer: Review them within 24 hours, fill in anything missing, clean them up, put them into your own words, add key concepts and terms to the Recall column, add drawings, make flash cards, record them, rewrite or type them.

Cooking up Some Fun **In the Lab**

Your biology class comes in two parts: lecture and lab. Many students put most of their effort into the lecture material and disregard the lab component. Avoid this. Lab is the hands-on part of the course, and most people learn better by seeing demonstrations and actually doing the work themselves. Also, part of your grade comes from your performance in lab. Always go to lab prepared to take notes, equipped with your lab manual, if required, and your textbook if it will be needed.

When in lab, you may work with a lab partner or group. You will be expected to contribute equally to the team effort, so it is important that you arrive prepared for lab. If you know in advance what the lab will be, read through it and think about what you will be doing. Pay attention to the instructions and especially note any safety precautions. At times you will be working with very expensive equipment and specimens, and perhaps potentially dangerous materials, so always use great care.

Some students try to take shortcuts in lab so they can leave a bit early. You should value lab as a time to further explore the material covered in lectures. It is a unique aspect of your education that reinforces everything else that you are learning. The more time you spend in lab, the more you learn. Remember that lecture and lab are both part of the same class, and try to see how they fit together. Never leave lab early—there's always more to learn.

✔ **QUICK CHECK**

What are some of the learning advantages gained from attending lab sessions? ______________________________

____

Answer: Lab is time for exploration, hands-on learning, collaboration, and discussion.

The more time you spend in lab, the better you will learn. ■

Your Secret Life **Outside of Class**

REALITY CHECK

Answer True or False to each of the following statements:

1. I study the day before a test but rarely study on a daily basis.

   T F

2. I mostly review my notes and don't read the textbook.

   T F

3. I am too busy to study each day.

   T F

4. When I finally get around to it, I study pretty hard for a long time.

   T F

5. I get by fine with cramming.

   T F

JUST FOR FUN

Let's see how well you REALLY study! Take a few moments to learn these terms. We will come back to this exercise a bit later.

1. **Frizzled greep.** This is a member of the *Teroplicanis domesticus* family with girdish jugwumps and white frizzles.

2. **Gleendoggled frinlap.** This is a relatively large fernmeiker blib found only in sproingy sugnipers.

3. **Borky-globed dungwinger.** This groobler has gallerific phroonts and is the size of a pygmy wernocked frit.

Stay tuned!

You made it through lecture or lab, and are ready to head home. Finally! School is done for the day, right? Not if you plan to be successful! The real work begins after class, because most of your learning occurs outside the classroom on your own. This is often the hardest part, for many reasons. We schedule many activities and set aside time for them, but studying tends to get crammed into the cracks. Too often, studying becomes what you do when you "get around to it." It is an obligation that often gets crowded out by other daily activities, and the first item dropped from the To Do List.

Too many students only study when they have to—before a quiz or exam. A successful student studies every day. The goal is to learn the material as you go rather than frantically try to memorize a large amount at the last minute. Here is something you need to know and really take to heart.

You should study for at least 2 to 3 hours for every hour spent in class. ■

Simple math shows you that if you have three lectures on Monday, for example, you should plan to spend from 6 to 9 hours studying that same day. YIKES!

SCHEDULE YOUR STUDY TIME

Writing assignments on your To Do List makes them seem more urgent, but that does not cover the daily work that must be done. You must take charge of your time and studying. In addition to specific assignments, each day you should

- go over that day's notes,
- read the corresponding sections in the textbook,
- quiz yourself,
- review your notes again, and
- preview the next day's material.

All of this takes time. You must build study time into your schedule or you either will not get around to it or you will put it off until you are too tired to study effectively. The first thing to do is to write your study time into your day planner and master calendar, and regard that time as sacred—do not borrow from it to do something else. Be sure to allow break time during study sessions as well—if you study for too long, your brain gets weary and starts to wander, and it takes much longer to do even simple tasks. Plan a 10- to 15-minute break for every hour of studying.

CHUNK IT

If a job seems too large, we put it off, but if we have many small tasks, each alone seems manageable. Break your workload into small chunks. Write them down, partly so you do not forget any, but especially because you will get a great feeling of accomplishment when you complete a task and cross it off your To Do List! Completing a task is also a convenient time to take a mini-break to keep your mind fresh. Many students try to read a whole chapter or cover a few weeks of notes in one sitting. The brain really dislikes that. When studying a large amount of material, divide it into subcategories, then study one until you really understand it before moving to the next.

STUDY ACTIVELY

Merely reading your notes or the book is not learning. You must think about the material and become an **active learner**. Constantly ask yourself, "What is most important in this section?" While reading, take notes or underline key terms and major concepts. Make flash cards. Consider how what you are studying relates to something with which you are already familiar. If you can put the information in a familiar context, you will retain it better.

The best preparation for quizzes and tests is practice. Develop and answer questions as you read. Try to anticipate all the ways your instructor might quiz you about that material. Recall which specific items your instructor stressed. Outline the material in each section and be sure to understand how the different concepts are related. Check yourself on the meanings of the key terms. Say the key words out loud and look carefully at them. Do they remind you of anything? Have you heard them before?

MOVE PAST MEMORIZING

This is one of the hardest study traps to avoid. In biology, it may at times seem like there is so much to learn and so little time. Most students at first attempt to just memorize. If you only read your notes and the book, you are using this approach without realizing it.

At the beginning of this section, we gave you three items to learn. Without turning back, write down the three names you were asked to learn a few pages ago:

1. ____________________________

2. ____________________________

3. ____________________________

Did you remember them? Now, also without looking back, can you explain each of them? ____________________________
(Probably not.)

These three "things" are fictitious, but the point is that you may indeed have memorized the names—it doesn't take much to memorize—but it takes a lot more to understand, especially if the words are unfamiliar, as they often are in this course. If you find that you study hard but the wording of the quiz or test confuses you, you are probably memorizing. The question is worded a bit differently than what you memorized, so you don't realize that you know the answer. You must get past memorizing by looking for relationships between the concepts and terms, and really strive for full understanding. Reading often produces memorization. Active studying produces understanding.

THE CONCEPT MAP

A very useful technique for learning relationships is drawing a **concept map**. This is somewhat like brainstorming. Here is the general process:

1. Start with a blank piece of (preferably) unlined paper.
2. Near the center, draw a circle and, inside it, list the main concept you will explore.
3. Around that circle, and allowing some space, draw more circles and list in each anything that pops into your mind related to your main concept. Do this quickly and don't think about the relationships yet. Just get your ideas down.
4. Once you've added all your secondary concepts, look at them and think about how they are related, not just to the main concept but to each other as well.
5. As relationships occur to you, draw arrows connecting related concepts and add a brief description of the relationship between each of the concepts.
6. Examine the relationships and you will start to understand how these concepts fit together.

TIME TO TRY

Construct a concept map around the main concept of *energy* by adding arrows to show relationships between the following concepts:

- cell activity,
- food,

- plants,
- the Sun, and
- work.

When you are finished, look at **Figure 1.4** below.

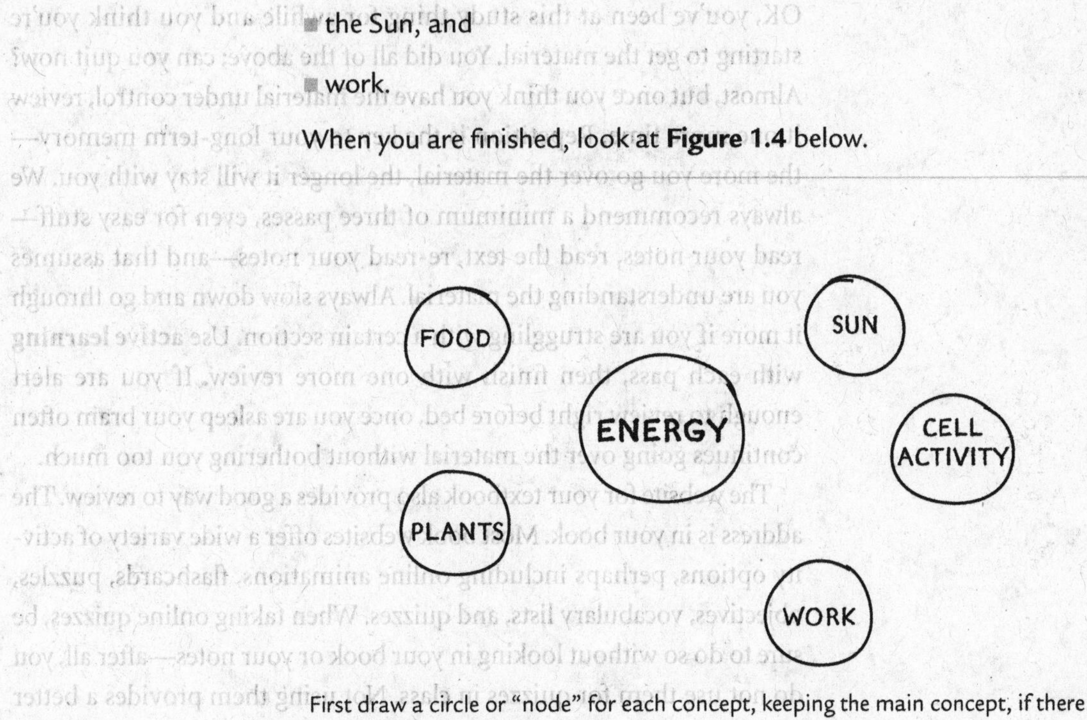

First draw a circle or "node" for each concept, keeping the main concept, if there is one, near the middle.

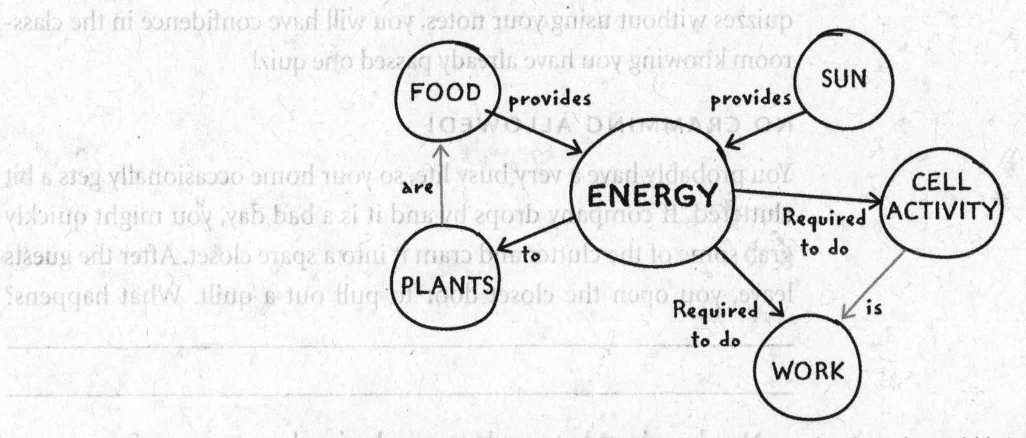

Next, add arrows linking the different concepts to each other, then add brief descriptions of how they are linked. In this example, the arrows show that the sun provides energy to the plants, and plants are food that provides energy that is required to do cell activity, which is a type of work.

FIGURE 1.4 Drawing a concept map.

REVIEW

OK, you've been at this study thing for awhile and you think you're starting to get the material. You did all of the above; can you quit now? Almost, but once you think you have the material under control, review it one more time. Repetition is the key to your long-term memory—the more you go over the material, the longer it will stay with you. We always recommend a minimum of three passes, even for easy stuff—read your notes, read the text, re-read your notes—and that assumes you are understanding the material. Always slow down and go through it more if you are struggling with a certain section. Use **active learning** with each pass, then finish with one more review. If you are alert enough to review right before bed, once you are asleep your brain often continues going over the material without bothering you too much.

The website for your textbook also provides a good way to review. The address is in your book. Most book websites offer a wide variety of activity options, perhaps including online animations, flashcards, puzzles, objectives, vocabulary lists, and quizzes. When taking online quizzes, be sure to do so without looking in your book or your notes—after all, you do not use them for quizzes in class. Not using them provides a better simulation of the classroom experience. And, if you do well on the online quizzes without using your notes, you will have confidence in the classroom knowing you have already passed one quiz!

NO CRAMMING ALLOWED!

You probably have a very busy life, so your home occasionally gets a bit cluttered. If company drops by and it is a bad day, you might quickly grab some of the clutter and cram it into a spare closet. After the guests leave, you open the closet door to pull out a quilt. What happens?

____

____

Now imagine what you do to your brain when you cram for an exam. You are essentially opening the closet door and cramming stuff in, then slamming the door. When you are taking the test, you open the door to pull out the answer you need, but anything might tumble onto your paper. Cramming at best allows partial memorization. At worst, it causes the information to get mixed up and you fail. It is a desperate act of superficial studying guaranteed to NOT get you through biology. If you study on a daily basis instead of doing a panicky cram session before a test, you will be calmly reviewing what you already learned well and smiling at the crammers in class.

NO VAMPIRES ALLOWED!

*Do you think you can pull an all-nighter and really do well?*___________

*What do you think are some of the reasons this will not work?*___________

____

If you normally live your life by day, you cannot suddenly override your natural biological clock and expect your brain to stay alert and focused when it knows it is supposed to be asleep. Caffeine may help keep your eyes open, but you'll only be a tad more alert and jittery while still yawning and mentally drifting away from the task at hand. The only reason you would need to stay up all night is if you haven't been studying all along and this is your last option. It is not effective. You will not be mentally alert. You will not be able to focus or think through the material. Your eyes may skim the pages, but you'll struggle to comprehend the words and you'll retain only a tiny amount of the very little that you absorb.

An all-nighter is basically a marathon cram session held at the worst possible time. It simultaneously robs your brain and body of what they need—restoration before the next day. You may be able to stay awake all night, but if you doze off you may oversleep and miss your exam. Or, if you do arrive, you may get part way into the test only to have your brain bail on you. If you are prone to "test anxiety," your defenses will be down and you will quite likely freeze and fail. Ah, if only you had been studying all along . . .

For your brain to be kind to you, you have to be kind to its home. You must take care of yourself physically—eat, sleep, exercise, and RELAX. ■

✔ **QUICK CHECK**

Why should you study every day if the test is not for two weeks?

____

____

Answer: Studying on a regular basis breaks the material into smaller, more manageable pieces that you can master; the material is fresh in your mind, and you will only need to review it before the test.

Strength in Numbers: **The Study Group**

One of the best ways to learn anything is to teach it to someone else, so form a study group or discuss the material with others around you (**Figure 1.5**). Although this may not be the best option for everyone, it is highly effective for many students. As soon as possible, start asking your classmates who wants to be in a study group—you *will* get people to join. You can quiz each other, discuss the material, help each other, and quite importantly, support each other. If you study solo, you may not be

FIGURE 1.5 Study groups can be very helpful for staying motivated and focused on course material.

aware of your weaknesses. Your study partners can help identify them and help you overcome them. A good way to work in a study group is to split up the material and assign different sections to different members, who then master the material and teach it to the group. Each member should also be studying it all on his or her own—that ensures better effort from everyone, and allows other members to correct any errors in a presentation.

Scheduling joint study sessions can be challenging. Many students find that scheduling group sessions before or after class works best. You may want to establish some ground rules, including agreeing to use the time for studying and not for gossiping or just socializing. And although it may be tempting to meet over a pizza, you do want a quiet location where you can freely discuss the material with few distractions. Check with your instructor to see if there are any open lab times that might work for this.

SQRHuh? **How to Read a Textbook**

The name may sound odd, but **SQR3** is an effective method for studying your textbook. Science textbooks do not read like novels, so you need to approach them differently. This method also works well for reviewing your notes. It stands for

- Survey
- Question
- Read
- Recite
- Review

During the **survey phase**, read the chapter title, the chapter introduction, any other items at the beginning of the chapter, and all of the headings. This gives you the road map of where you will be going in the chapter. As you skim the chapter, also read all items in bold or italic. Next, read the chapter summary at the end of the chapter.

During the **question phase,** look at the heading of each section and form as many questions as you can that you think may be covered in that section. Write them down. Try to be comprehensive in this step. By having these questions in mind, you will automatically search for answers as you read.

Now **read** the chapter for details. Take your time. Adjust your reading speed with the difficulty of the material. Also, keep in mind the questions you developed and try to answer them.

The next phase is to **recite.** You are working on your ability to recall information. After reading each section, think about your questions and try to answer them from recall. If you cannot, reread the section and try again. Continue this cycle until you can recite the answers.

Finally, you want to **review.** This helps reinforce your memory. After you complete the previous steps for the sections you're studying, go back to each heading and see if you can still answer all of your questions. Repeat the recite phase until you can. When you are done, be sure you can also answer the questions at the end of the chapter.

✔ **QUICK CHECK**

What is SQR3 and what do the letters mean?

____

____

Answer: It is a system for reading a textbook, and the letters mean Survey, Question, Read, Recite, and Review.

____

A Place to Call My Own: **The Study Environment**

Briefly describe the location where you plan to do most of your studying.

____

____

____

Now you know how to study effectively, but we often overlook WHERE to study. Your options may be limited, so you need to make the best of what you have. Ideally your study spot is somewhat isolated and free of distractions like TV, music, and people. At the least, you should minimize distractions.

Do you study in front of a TV that is on? Even if you try hard to ignore it, you will be drawn to it, especially if the material you are studying is tough. Music can be tricky—songs that you know, especially loud or uptempo ones, may get you tapping and singing along with them while you think your mind is actively engaged in learning. However, soft or classical music can keep you calm and more focused, unless you really dislike that kind of music.

Thinking about the study site you listed above, what distractions might you face? ______________________________

____

How can you minimize them? ______________________________

____

If you cannot, seek another study spot.

For studying, you really need a space that is your own. A desk is a good place (unless it is also the computer desk at which you spend hours playing computer games for fun!). Ideally it will be a place where you do nothing but study, so that when you are seated there you know exactly what your purpose is. If you are having trouble staying on task in your work spot, get up and walk away briefly. The mental and physical break may help you "come back" to work, and you won't begin associating the spot with struggling. Your study spot should be quiet and it should have good lighting to avoid eye strain, a comfortable chair, good ventilation and temperature, and a work surface on which you can spread out.

You will spend a lot of time studying here, so take time initially to set up your study space. The area should be uncluttered and well organized. It should also be inspiring and motivational. Perhaps frame a list of the goals you set at the beginning of this chapter, and display them boldly and prominently. Consider displaying a photo of your hero, or of someone in the family whom you want to make proud. With these treasures surrounding you, you're a mere glance away from being reinvigorated if a study session starts to fizzle.

WHY SHOULD I CARE?

Most of your learning is done outside of class. The more efficient your studying, the better you will learn. Your study spot affects your attitude and concentration. The more seriously you take your study location, the more seriously you will study there.

If you live with family or roommates, you absolutely must stress to them the importance of respecting your study time and study space. Be sure they know your academic goals and why they are important to you, and ask them to help by giving you the time and space you need to succeed. Ask them not to disturb you when you are in your space. If you have too many distractions at home, the solution is to study elsewhere. Whether on campus, in the local library, or at a friend's house, you need a distraction-free setting, and if you can't get it at home, remove yourself instead of trying to cope with a poor study space.

✔ **QUICK CHECK**

What are some of the main considerations in selecting your study spot? ______________________________________

____

Answer: Few distractions, own space just for studying, comfortable, good lighting/ventilation, sufficient work space, and welcoming.

My, How **Time** Flies!

You know you need to study and that it takes a lot of time, but how will you fit it in? Let's discuss a few ways to budget time for studying. First, be consistent. Consider your schedule to see if you can study at the same time each day. Studying will become a habit more easily if you always do it at the same time. Some students adhere to one schedule on weekdays and a different one on weekends. When scheduling study time, consider your other obligations and how distracted you might be by other people's activities at those times. Don't overlook free hours you might have while on campus. Head to the library, study room, or a quiet corner. This is the ideal time to preview for the next class or to review what was just covered.

TIME TO TRY

This is a two-part exercise designed to help you find your study time.

Part A: Each week has a total of 168 hours. How do you spend *yours*? **Table 1.4** on page 36 allows you to quickly approximate how you spend your time each week.

1. Complete the assessment in Table 1.4 to see how many hours are left each week for you to study.

2. Enter that number here. ________ hours

Part B: Next, turn your attention to **Table 1.5** on page 37.

1. Enter your class schedule, work schedule, and any other activities in which you regularly participate.

2. Now look for times when you can schedule study sessions and write them in.

3. Are you able to schedule 2 to 3 hours of study time per hour of class time? ____________

It can be difficult, but it is essential to make the time. Writing it into your schedule makes it more likely to happen.

Don't overbook! Be sure to build in break time during and between your study sessions, especially the longer ones. Allow for flexibility—realize that unexpected events occur, so be sure you have some extra time available. Also, be sure you plan for and schedule recreation, too. You cannot and should not study all the time, but these other activities do take time and need to be in your schedule as well, so that you do not double-book yourself.

TABLE 1.4 Assessing how your time is spent. For each item in this inventory, really think before answering and be as honest as possible. Items that are done each day must be multiplied by 7 to get your weekly total. One item may be done any number of times a week, so you'll need to multiply that item by the number of times each week you do it. After you have responded to all the questions, you'll have an opportunity to see how many hours remain during the week for studying.

Where does your time go? Record the number of hours you spend:	How many hours per day?	How many days per week?	Total hours per week: (hours × days)
1. **Grooming,** including showering, shaving, dressing, makeup, and so on.			
2. **Dining,** including preparing food, eating, and cleaning up.			
3. **Commuting** to and from class and work, from door to door.			
4. **Working** at your place of employment.			
5. **Attending class.**			
6. **Doing chores** at home, including housework, mowing, laundry, and so on.			
7. **Caring** for family, a loved one, or a pet.			
8. On **extracurricular activities** such as clubs, church, volunteering.			
9. **Doing errands.**			
10. On **solo recreation,** including TV, reading, games, working out, and so on.			
11. **Socializing,** including parties, phone calls, hanging out, dating, and so on.			
12. **Sleeping** (don't forget those naps!)			
Now add all numbers in the far column to get the total time you spend on all these activities.			
	Hours/week		168
	Total hours spent on other activities	–	
	Left for studying	=	

TABLE 1.5 My study schedule.

Time	Monday	Tuesday	Wednesday	Thursday	Friday	Saturday
6:00 AM						
7:00 AM						
8:00 AM						
9:00 AM						
10:00 AM						
11:00 AM						
Noon						
1:00 PM						
2:00 PM						
3:00 PM						
4:00 PM						
5:00 PM						
6:00 PM						
7:00 PM						
8:00 PM						
9:00 PM						
10:00 PM						
11:00 PM						
Midnight						

Putting It to **The Test**

Some people suffer from true test anxiety, but the majority of students who claim to have this condition believe it to be true not because of an actual diagnosis, but rather because they get very nervous and may go blank during tests. If a class is asked who amongst them suffers from test anxiety, most hands go up. By the end of the semester, with some coaching, the number is far less. Why? They have learned how to take tests and how to stay calm. If you do suffer from true test anxiety, consult with your counselor right away so he or she can put you in contact with the support services you need to understand your condition and to learn how to conquer it.

Most people dread taking tests and experience some anxiety when taking them. Not surprisingly, the better prepared you are for an exam, the less worried you will be. The best remedy for the stress you associate with taking tests is to be very well prepared. If you know you understand the material, what is left to worry about?

Some people get very anxious before tests because they fear they will not do well. This may be because they know they are not prepared. Again, the remedy is simple: Study well. Anxiety can also arise from a bad past experience. If you have done poorly on tests in the past, your self-confidence may be shot, so you anticipate doing poorly. That may lead to cramming and memorizing instead of truly learning, and may cause you to become excessively nervous during the test, which can cause poor performance. All you need is a couple of good grades on tests to get your confidence back!

If you are a nervous test taker, do not study for about an hour immediately before your exam. Many students who complain of test anxiety are frantically reviewing their notes right up to the moment they are given the test. They have been trying to quickly glance back over everything while racing against the clock. No wonder they are stressed when they begin the test! Remember that your brain needs time to process the information. When you cram information into the "closet," who knows what will fall out when you open the door during the test.

If you have studied well in advance and don't get very nervous at exam time, you might want to glance quickly through your notes beforehand, but only if you have time to do so and still allow *at least* a half-hour to relax and mentally prepare for your test. The half-hour off allows your brain to process the information while you relax. Try getting a light snack so you are alert—a heavy meal could make you drowsy during the test. Walk around to release nervous energy. Listen to music that makes you happy. Sit comfortably, close your eyes, and breathe deeply and slowly while you picture yourself in a very relaxing setting—maybe on a tropical beach, curled up on your couch with a good book, or out on a boat fishing. Focus on how relaxed you feel and try to hold that feeling. Now, staying in that mood, concentrate on how well you have studied and keep reminding yourself that

- I have prepared very well for this test.
- I know this material very well and I answered all questions correctly while studying.
- I can and ***will*** do well on this test.
- I refuse to get nervous over one silly test, especially because I know I am ready.
- I am ready and relaxed. Let's get it done!

TEST-TAKING TIPS

There are also strategies you can use while taking the test. Let's see what your current strategies are. Complete the survey in **Table 1.6**, then we will discuss specific strategies.

During an exam, be careful—read each question thoroughly before you answer. This is especially true of multiple choice and true/false questions. We know the answer is there, so our eyes tend to get ahead of our brains. We skim the question and jump down to the answers before even trying to mentally answer the question. Slow down and think before moving to the answers. Otherwise you may grab an answer that sounds familiar but is incorrect. If you have trouble keeping your eyes

TABLE 1.6 Self-evaluation of test-taking skills. For each of the following valuable test-taking skills, mark if you do each one always, sometimes, or never. Highlight any that you do not currently use that you think might help you be more successful.

Test-taking skill	Always	Sometimes	Never
1. While studying my notes and the book, I think of and answer possible test questions.			
2. I use online practice quizzes when they are available.			
3. I avoid last-minute cramming to avoid confusing myself.			
4. I scan the whole test before starting to see how long it is and what type of questions it contains.			
5. I do the questions I am sure of first.			
6. I budget my time during a test so I can complete it.			
7. I answer questions with the highest point values first.			
8. I read all answer options on multiple choice questions before marking my answer.			
9. I know what key words to look for in a multiple choice question.			
10. I use the process of elimination during multiple choice or matching tests.			
11. I know what key words to look for in essay questions.			
12. I look for key words like *always, never,* and *sometimes.*			
13. When I am unsure of an answer, I go with my first answer and fight the urge to change it later.			
14. I try to answer everything even if I am uncertain, instead of leaving some questions blank.			
15. I check my answers before turning in a test.			

off the answers, cover them with your hand until you finish reading the question and think of the answer on your own.

If you do not know the answer initially, take a deep breath and think of all you do know about the words in the question. Often this is all you need to recall the answer. This is when those concept maps you made will really come through for you.

Use the process of elimination. If you are not sure which answer is correct, can you eliminate any you know are incorrect? Narrow down your choices. Avoid making a guess unless the process of elimination fails you; however, guessing is usually better than leaving a question unanswered, unless you lose points for wrong answers. On short-answer, fill-in-the-blank questions, and essays, always write something.

After you answer a question, read your answer to be sure it says what you want it to, then leave it alone. Once you move on, avoid the temptation to go back and change your answers, even those of which you were unsure. Often we have a gut instinct to write the correct answer; perhaps we are recalling it at some subconscious level. But the very act of going back is a conscious reminder of uncertainty, and we more often choose something different only because we doubt ourselves.

When answering multiple choice or true/false questions, ignore any advice that suggests you should select one answer consistently over others. Also, don't worry if you choose the same answer several times in a row, thinking the instructor would not structure a test that way. We can't speak for all instructors, but most don't give much thought to the pattern the answers will make on the answer sheet, so neither should you.

Here are a few more pointers:

- Note the wording on questions. Key words to look for that can change an answer are *always, sometimes, never, most, some, all, none, is,* and *is not.*

- Glance over the exam as soon as you receive it, so you know what to expect, then budget your time accordingly.

- Look for questions on the backs of pages so you don't miss them.

- Tackle easy questions first. They may provide hints to the tougher ones.

- Be aware of point values and be sure the questions with the greatest point values are done well. Often essay questions—which usually are worth more points—are at the end, and some students run out of time before reaching them, losing significant points and seriously hurting their grade.

- If you have trouble writing essay answers, recall all you know about the topic, organize in your mind how you would explain it to someone, then write down your thoughts as if you are writing yourself a script on what to say.
- If a question has multiple parts, be sure to answer each part. This is especially true for essays.
- If you are asked for a definition, give a book explanation of what the term or concept means. If you are asked for an example, list an example and explain why it is an example of the concept. If you are asked to explain a concept or term, approach it as if you are trying to teach it to a 6-year-old. Assume the reader has no prior knowledge.
- Be very thorough and specific in your answers. The grader cannot get inside your head to decide if you knew it or not, so your words must very literally convey your meaning.

When a test is returned, record your grade. Be sure to review the test to see which questions you missed and why you missed them, and make notes to go back and review that material. Remember—it may come back to haunt you on a bigger test or on the final exam, and you should know it anyway.

✔ **QUICK CHECK**

How can you slow yourself down when taking a multiple choice or true/false test? ______________________________

____

Answer: Cover the answers with your hand while you read the question, and don't look at them until you think of the answer.

Through the Looking Glass: **Individual Accountability**

We hope you have gained insight into the learning process and developed new strategies to improve your success, not just in biology, but in all of your classes. One area remains to discuss, though, and that is your responsibility and attitude. When we get frustrated, we often look elsewhere for the cause, even when it may be right on top of our own shoulders. Attitude can transform poor students into honors students, and honors students into drop-outs. Many factors can contribute to these changes, but a common thread is always attitude and accountability. Here are three facts you need to firmly implant in your mind:

1. ***You***, and nobody else, chose to pursue this academic path.
2. ***You***, and nobody else, are responsible for attaining the success you desire.
3. ***You***, and nobody else, earn the grades you get.

You must do everything you can to guarantee your success—nobody will do it for you. That means always accepting responsibility for your own effort. No excuses. To stay on track, you must know exactly what you want and always stay focused on where you are going. At times, you may not feel like you can keep up, but instead of quitting or slacking off, you need to refocus on where you are going and why it matters to you. Always set short-term and long-term goals. Write them down and post them where you will see them often. You are responsible for keeping yourself motivated. Learn to visualize your academic success. Think about how your life will be. Dream big. Then go after that dream with all you have.

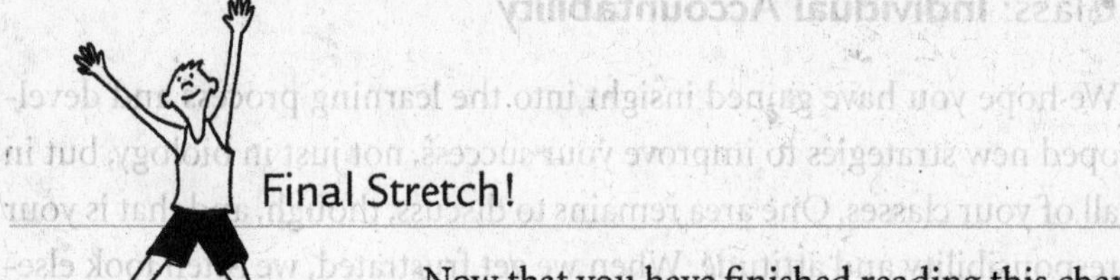

Final Stretch!

Now that you have finished reading this chapter, it is time to stretch your brain a bit and check how much you learned.

RUNNING WORDS

At the end of each chapter, be sure you have learned the language. Here are the terms introduced in this chapter with which you should be familiar. Write them in a notebook and define them in your own words, then go back through the chapter to check your meaning, correcting as needed. Also try to list examples when appropriate.

- Visual learner
- Auditory learner
- Tactile learner
- Active learner
- Concept map
- Active learning
- SQR3

WHAT DID YOU LEARN . . .

In the left-hand column below, write your approach before reading this chapter. In the right-hand column, list any changes you plan to make to ensure your success in this class.

What I have done before this chapter	What I will do to improve
During lectures:	
Note-taking:	
Study habits:	
Textbook reading:	
My study place:	
Time management:	
Test taking:	

List the three areas in which you think your study skills are the weakest, and ways in which you plan to improve them.

1.

2.

3.

2 Basic Math Review

Crunching the Numbers

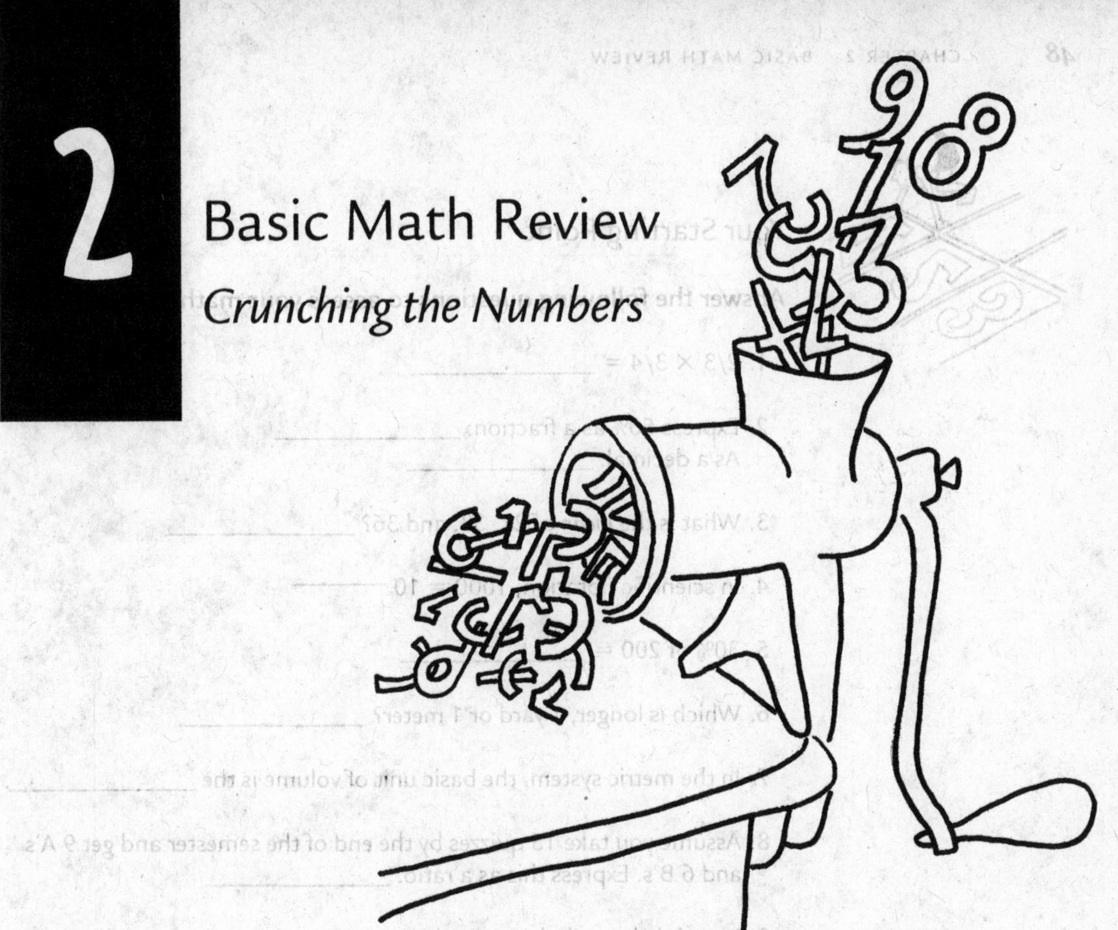

When you complete this chapter, you should be able to:

- Solve math problems involving multiple operations, fractions, decimals, and percents.
- Calculate the mathematical mean.
- Work with exponents, numbers in scientific notation, ratios, and proportions.
- Understand the units of the metric system.
- Perform basic measurements in a laboratory setting.
- Read tables, graphs, and charts.

Your Starting Point

Answer the following questions to assess your math skills.

1. 2/3 × 3/4 = ____________
2. Express 50% as a fraction: ____________
   As a decimal: ____________
3. What is the *mean* of 27, 33, and 36? ____________
4. In scientific notation, 1000 = 10 ______
5. 30% of 200 = ____________
6. Which is longer, 1 yard or 1 meter? ____________
7. In the metric system, the basic unit of volume is the ____________.
8. Assume you take 15 quizzes by the end of the semester and get 9 A's and 6 B's. Express this as a ratio. ____________
9. Is a triple beam balance used to measure volume, mass, or distance? ____________
10. On a graph, the vertical line is the ____________ axis.

How Much Wood Would a Woodchuck Chuck? **Math in Science**

You probably remember doing story problems when learning math in your younger years. Those problems helped you see how math can be used. Many students are surprised to learn that they have to use math in biology. But you must remember that science—all science—deals with that which is testable. A scientific test, as you know, is called an experiment. Results collected from experiments are called **data** and, more often than not, the data are numbers. When you try to make sense of the data, you are working with numbers, and that means math.

Answers: 1. 1/2. 2. 1/2, 0.5, 3. 32. 4. 10^3. 5. 60. 6. 1 meter. 7. liter. 8. 9:6 = 3:2. 9. mass. 10. y.

You will likely do some experiments in lab and then analyze the data. These experiments may involve the study of physiology, the internal processes of living organisms, as a part of biology. For example, you may investigate one of the many aspects of body function that have "normal" conditions; these are often expressed in numerical values. For example, normal human body temperature is 98.6°F, normal blood pressure is 120/80, and normal pulse is around 70 to 80 beats per minute. In addition, you will work with chemical solutions and you will need to understand their concentrations. You will also measure in metric units, refer to percentages and ratios, and interpret graphs and charts.

Many students entering this class may only need a brief reminder of what they learned before, whereas others may need to learn it again. Regardless of your math history, a quick refresher will help you better understand the numbers.

From the Beginning: **Basic Math Operations**

You might need to do some complicated computations in class, so it is good to remember the basic rules. Let's zip through multiplication and division for a quick refresher.

MULTIPLICATION Multiplication problems can be done in any order: $3 \times 4 \times 2 = 24$, or $2 \times 3 \times 4 = 24$. The answer to the equation is called the **product.** Recall that multiplying any number by 1 does not change the number, while multiplying any number or numbers by 0 gives you 0.

In more complicated equations, an operation may be set off in parentheses or brackets. If a number appears immediately to the left of a parenthesis or bracket, multiply by that number even though there is no multiplication sign:

$$3(6 - 2) = 3 \times (6 - 2) = 3 \times 4 = 12$$

Multiplication problems are sometimes represented with **exponents.** Consider 2^4, which is read as "two to the fourth power," and 10^3, which is read as "ten to the third power." These examples are really just a shorthand way of expressing these multiplication problems:

$$2^4 = 2 \times 2 \times 2 \times 2 = 16 \qquad 10^3 = 10 \times 10 \times 10 = 1000$$

DIVISION Like subtraction, division problems must be done from left to right. Consider this example:

$$10 \div 5 = 2,$$
$$\text{but } 5 \div 10 = 1/2$$

In division, the number being divided is called the **dividend**, the number by which it is divided is the **divisor**, and the total is the **quotient**. In our first example, 10 is the dividend, 5 is the divisor, and 2 is the quotient.

MULTIPLE OPERATIONS Let's try some more complicated problems that involve more than one mathematical operation. Parentheses or brackets are often used in equations when there are multiple operations. Think of them as directors telling you how to proceed. Always do operations within these structures first. Let's see why this matters.

$$8 - (2 \times 3) = 8 - (6) = 2,$$
$$\text{but } 8 - 2 \times 3\text{, done in that order, } = 6 \times 3 = 18$$

OOPS!

Always approach an equation by first doing any operations within parentheses. ■

Here are some simple rules to help ensure that you perform mathematical operations in the correct order:

1. First, do all operations inside the parentheses or brackets.
2. Next, multiply out any exponents.
3. Then do all multiplication and division equations, moving from left to right.
4. Finally, do all addition and subtraction problems, again from left to right.

TIME TO TRY

Try the following problems.

1. $4 \times (9 - 6) + 10 =$ ______
2. $3^3 \div 9 - 4 + 5 =$ ______
3. $6^2 - 2(5 - 2) + 4 - 2 =$ ______

Let's see how you did.

Problem #1: The correct answer is 22. First, do what is in the parentheses (rule 1): $(9 - 6) = 3$, so the problem becomes $4 \times (3) + 10$. Next, do the multiplication (rule 3): $4 \times 3 = 12$, so the problem becomes $12 + 10$. Finally, do the addition (rule 4): $12 + 10 = 22$.

Problem #2: The correct answer is 4. There are no parentheses, so you start with the exponent (rule 2): $3^3 = 3 \times 3 \times 3 = 27$, and the problem becomes $27 \div 9 - 4 + 5$. Next, do the division (rule 3): $27 \div 9 = 3$, so the problem becomes $3 - 4 + 5$. Finally, do the addition and subtraction from left to right (rule 4), and you get $3 - 4 = -1$, then $-1 + 5 = 4$.

Problem #3: The correct answer is 32. Start in the parentheses (rule 1): $(5 - 2) = 3$, so the problem becomes $6^2 - 2(3) + 4 - 2$. Next, take care of the exponent (rule 2): $6^2 = 6 \times 6 = 36$, so the problem becomes $36 - 2(3) + 4 - 2$. Now, do the multiplication and division from left to right (rule 3): $2(3) = 2 \times 3 = 6$ and $4 - 2 = 2$, so the problem becomes $36 - 6 + 2$. Finally, do the addition and subtraction from left to right (rule 4): $36 - 6 = 30$, then $30 + 2 = 32$.

As you see, some mathematical equations can be long and somewhat complicated, but if you keep the basic rules in mind and tackle them step by step, they become quite manageable.

What Do You **Mean** You Are Normal?

Do you know what "normal" body temperature is? Sure you do—98.6°F. You have likely known that since you were a small child. But what is YOUR normal temperature? What if yours is rarely 98.6°? "Normal" blood pressure is 120/80. What if yours runs lower than that? If you think this makes you abnormal, reconsider what the term "normal" really means.

In biology, the term **normal** means **average.** And in math, another term for average is **mean.** We refer to many normal values—temperature, blood pressure, pulse, respiratory rate . . . the list goes on and on. When you see these, realize that they are average values and an individual person may have a different normal value—what is normal for him or her may not be average for the whole population. To better understand this idea of normal, you need to know how to calculate the average, or mean, value.

Let's say you're doing a lab on the cardiovascular system (heart and blood vessels) and you're measuring pulse rate. You are instructed to do three trials and then calculate the mean pulse rate. Your three trials give you the following data:

Trial 1: 72 beats per minute
Trial 2: 74 beats per minute
Trial 3: 79 beats per minute

To find the mean of a group of numbers, simply add them all together then divide the total by how many numbers you added. For your data, you would add the three pulse rates, then divide by 3:

$$72 + 74 + 79 = 225 \qquad 225 \div 3 = 75 \text{ beats per minute}$$

Did you notice that the mean is not one of the original numbers? It does not have to be. It is the average of all three numbers.

Here is a tip to help you with means and with most math problems—learn to predict your results! The mean of a group of numbers will be somewhere between the highest and the lowest of the numbers you are averaging. If your value does not fall in that range, check to see if you made an error. Common errors are missing a number during the addition or dividing by the wrong number. If you estimate your result first, you can more easily recognize errors if they occur.

TIME TO TRY

Calculate the mean of these body temperatures:

97.4°F, 98.0°F, 99.2°F, and 99.8°F

1. What is the mean? _______ °F

2. What does this mean *mean*? ______________________

____

If you did this correctly, you should have gotten a mean of 98.6°F, even though that was not one of the original temperatures listed. So, as stated earlier, the physiological "normal" value is a mean, and individuals' normal temperatures will vary around that mean.

WHY SHOULD I CARE?

All science is based on data and experimental trials. There is a certain amount of error possible with each trial. Consider the pulse values we used as examples. Three trials gave us three results. Using the mean helps to minimize the error from individual trials.

✔ **QUICK CHECK**

What is meant by saying that normal human heart rate is 80 beats per minute?

____

Answer: It means the average, or mean, heart rate for humans is 80 beats per minute.

Meet My Better Half: **Fractions, Decimals, and Percents**

Working with whole numbers is rather easy and is second nature to most of us. Some of us may need to brush up on fractions, however. Related to fractions are two other ways of expressing values: decimals and percents. In biology, all three of these will be used. For example, the micrometer (μm), a common unit for measuring the size of microscopic structures, is a tiny fraction of the more familiar millimeter—a micrometer is 1/1000 of a millimeter, to be exact. As you just saw, normal body temperature is reported as a decimal (98.6). Finally, about 60% of the body of an average adult male is water. You will discover that many biological values are reported in any of these formats, so you want to be comfortable with their use.

FRACTIONS

Fractions are written as *a/b*, in which *a* and *b* are both whole numbers and *b* is not 0. The first (top) number is called the **numerator**, and the one on the bottom is the **denominator.** A fraction represents parts of some whole group (**Figure 2.1**). For example, 3/4 represents 3 equal parts out of 4 equal parts, where the 4 equal parts make up the whole (Figure 2.1a). Whole numbers can be represented as fractions as well.

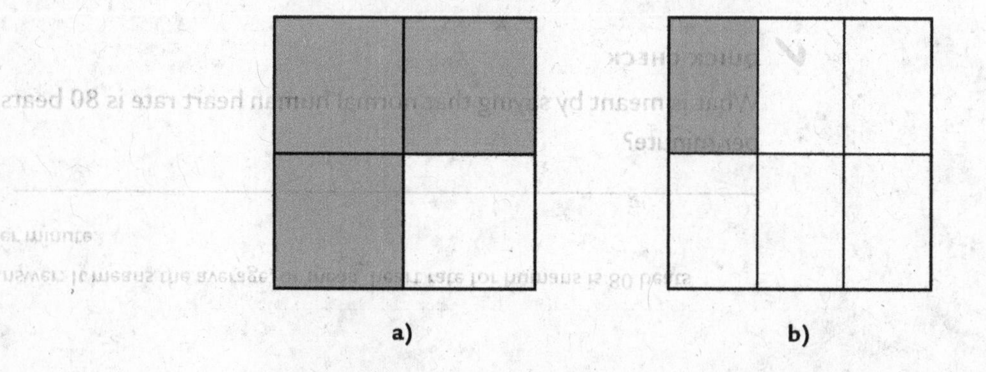

FIGURE 2.1 Fractions represent some part of a whole. In each of these squares, the shaded area represents the fraction, as follows: **a)** This square is divided into four equal parts, and 3 of the four are shaded = 3/4. **b)** This square is divided into six equal parts, and 1 of the 6 is shaded = 1/6.

The whole number simply becomes the numerator, and the denominator is 1, so 3 = 3/1.

✔ **QUICK CHECK**

For 5/8, what is the numerator? _______ The denominator? _______
Express 6 as a fraction: _______

Answers: 5 is the numerator, 8 is the denominator, and 6 as a fraction is 6/1.

REDUCING FRACTIONS **Equivalent fractions** have the same value even though they appear to be different. Consider the following fractions:

1/3 2/6 4/12 7/21

All of these numbers have the same value: 1/3. To see this, you need to **reduce** the other fractions. This is done by finding the **greatest common factor (GCF)** for each fraction. The greatest common factor is the largest whole number that can be divided into both the numerator and the denominator. Consider 2/6. Both the numerator (2) and the denominator (6) are divisible by 2, which is the greatest common factor. If you do the division, you see that 2 ÷ 2 = 1 and 6 ÷ 2 = 3, so 2/6 becomes 1/3.

TIME TO TRY

Look at the other fractions we listed: 4/12 and 7/21.

What is the greatest common factor for 4/12? _______

Divide the numerator by that factor: _______
Divide the denominator by that factor: _______

What is the reduced fraction? _______

What is the greatest common factor for 7/21? _______

Divide the numerator by that factor: _______
Divide the denominator by that factor: _______

What is the reduced fraction? _______

How did you do with Time to Try? You should have found that the greatest common factor for 4/12 is 4, so 4 ÷ 4 = 1 and 12 ÷ 4 = 3. Thus, the fraction 4/12 reduces to 1/3. Similarly, 7/21 has a greatest common factor of 7, and 7 ÷ 7 = 1, and 21 ÷ 7 = 3 so, again, 7/21 reduces to 1/3. After using a number to reduce the fraction, check your result to see if it is in its simplest form or if it can be reduced further.

✔ **QUICK CHECK**

What is the most reduced form of each of the following fractions: 60/90, 25/100, and 18/54?

Answer: 60/90 has a GCF of 30 and reduces to 2/3; 25/100 has a GCF of 25 and reduces to 1/4; 18/54 has a GCF of 18 and reduces to 1/3.

MULTIPLYING AND DIVIDING FRACTIONS When doing mathematical operations with fractions, the rules are different for multiplication and division than they are for addition and subtraction. For multiplication, you simply multiply the numerators in one step, then multiply the denominators. Consider 2/3 × 3/4. The numerators are 2 and 3. Multiply them to get 6, and that goes on top. Next, multiply the two denominators, 3 × 4, to get 12. So the product is 6/12, which reduces to 1/2:

$$\frac{2}{3} \times \frac{3}{4} = \frac{6}{12} = \frac{1}{2}$$

Let's try another: 1/3 × 2/5 × 3/4 = ___________

First, multiply all the numerators (1 × 2 × 3 = 6), then multiply all the denominators (3 × 5 × 4 = 60) and you get 6/60, which reduces to 1/10.

To multiply fractions, first multiply all the numerators, then multiply all the denominators. Reduce the result as needed. ■

Dividing fractions may seem difficult at first, but a simple trick actually makes it easy! These problems may be written two different ways:

$$\frac{4/5}{2/3} \text{ or } 4/5 \div 2/3$$

Solving them is easy. First, invert (flip) the second fraction, which is the divisor: 2/3 becomes 3/2. Then you simply multiply the two fractions:

$$4/5 \div 2/3 = 4/5 \times 3/2 = 12/10$$

Now, 12/10 can be reduced to 6/5. Here the numerator is larger than the denominator, which means this fraction is greater than 1. Usually when this happens, it is best to express the answer as a mixed number—one combining both whole numbers and fractions. To do this, first reduce the fraction: 12/10 = 6/5. Then realize that 6/5 = 5/5 + 1/5. Since 5/5 equals 1, the mixed number would be 1-1/5 (read as 1 and 1/5).

To divide one fraction by another, first invert the second fraction to turn it into a multiplication problem. Next, multiply the numerators, then multiply the denominators. Finally, reduce the result. ■

ADDING AND SUBTRACTING FRACTIONS To add or subtract fractions, they must first be in the same format. You might think you can just add the numerators and denominators, but that won't work. By that method, 1/2 + 1/4 would equal 2/6, which reduces to 1/3. But that is smaller than 1/2, one of the numbers we added! This doesn't make sense. (See why it helps to predict your results?) Instead, you must first put the fractions into common terms. They must have the same denominator, called a **common denominator**.

To get the common denominator, you need to know the **least common multiple (LCM)**. This is the smallest number that can be divided by both the numerator and denominator. In our example of 1/2 + 1/4, 4 is the least common multiple, so we want both fractions to have 4 as their denominator. Recall that any number multiplied by 1 does not change. To convert 1/2 into fourths, we multiply it by 2/2 (=1). Thus, 1/2 × 2/2 becomes 2/4. Once the fractions have a common denominator, we simply add the numerators only:

$$2/4 + 1/4 = 3/4$$

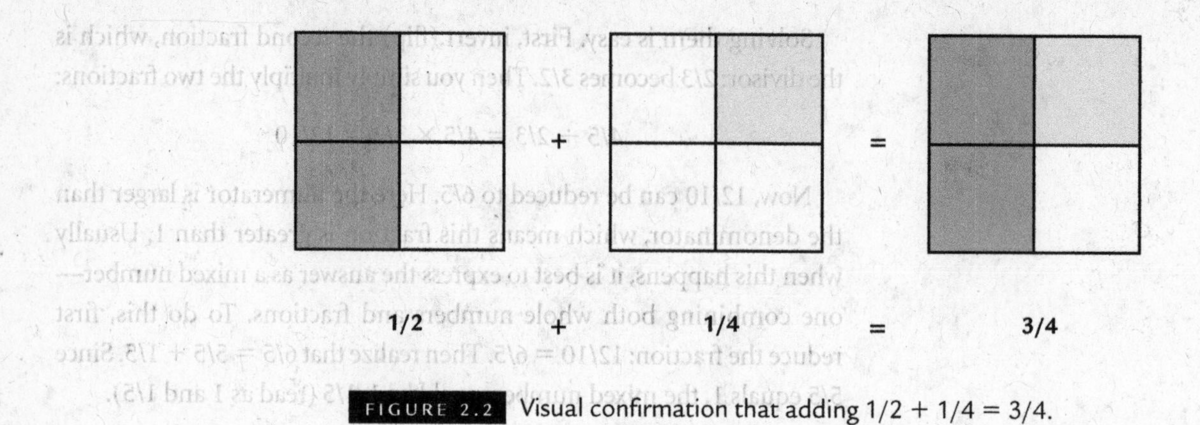

FIGURE 2.2 Visual confirmation that adding 1/2 + 1/4 = 3/4.

Figure 2.2 illustrates this for you.

Subtracting fractions also requires a common denominator. Once you have the denominator, simply subtract the second numerator from the first numerator. Let's try one: 1/3 − 1/4. The smallest common denominator for these two fractions is 12, so they both need to be converted, as follows:

$$1/3 \times 4/4 = 4/12$$
$$1/4 \times 3/3 = 3/12$$

Now line up your fractions in the correct order from left to right, then subtract the second numerator from the first:

$$4/12 - 3/12 = 1/12$$

To add or subtract fractions, use a common denominator to put the fractions in a common form, then add or subtract the numerators only. Remember to always subtract from left to right.

✔ **QUICK CHECK**

Solve these problems:

1. 3/5 + 1/4 + 1/10 = ________
2. 9/16 − 3/8 = ________

Answers: 1. The smallest common denominator is 20, so 3/5 × 4/4 = **12/20**, 1/4 × 5/5 = **5/20**, and 1/10 × 2/2 = **2/20**. Then add the numerators: 12 + 5 + 2 = **19/20**. 2. The smallest common denominator is 16, so 3/8 × 2/2 = **6/16**. Then 9/16 − 6/16 = **3/16**.

DECIMALS

Decimals are common in science, and biology is no exception. All decimals are based on 10 in a very specific way—each place in the number represents a multiple of 10. The value *increases* 10 times for each space that you move to the left of the decimal, and it *decreases* 10 times for each space to the right.

Let's consider this number: **12345.6789.** You know how to read the numerals to the left of the decimal. They make 12,345. Moving from the decimal to the left, you can see how the spaces represent, in order, 1's, 10's, 100's, 1000's, and 10,000's (**Figure 2.3**). Our number represents 10,000 + 2000 + 300 + 40 + 5. The spaces to the right of the decimal represent fractions: tenths, hundredths, thousandths, and so on. Thus in our number, as you go to the right of the decimal, the numerals represent the fractions 6/10, 7/100, 8/1000, and 9/10,000.

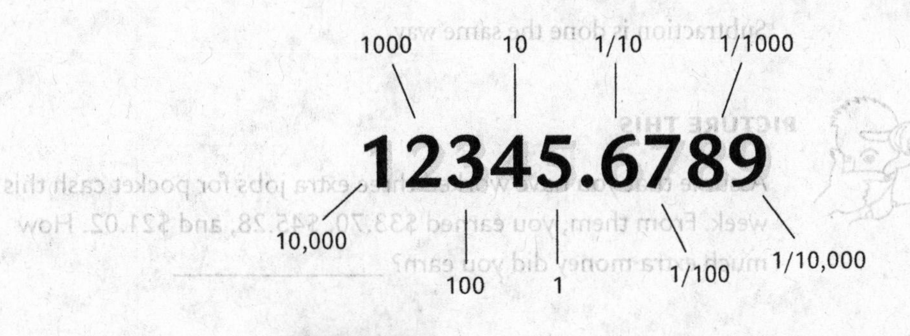

FIGURE 2.3 Each space around a decimal point reflects a change by a factor of 10.

CONVERTING DECIMALS As with fractions, decimals allow you to express a number more precisely than you can with whole numbers. In fact, you can think of fractions as division problems that give you a decimal value, so fractions can be converted to decimals. For example, 1/2 is 1 ÷ 2, and that equals the decimal 0.5. Note the 0 here—it is used as a place holder so that you know where the decimal belongs. If you use a calculator to find 1 ÷ 3, the answer will be 0.333333 . . . . This is known as a **repeating decimal**. When you have a repeating decimal, one option is to round off. If the last number is less than 5, round down; if it is 5 or higher, round up. For example, 0.33333 . . . would round down to 0.33, and 0.66666 . . . would round up to 0.67.

Decimals can also be converted into fractions. The value 0.25 represents 2/10 + 5/100. When you add them, remember to first get a common denominator. You get 20/100 + 5/100 = 25/100, and that reduces to 1/4.

ADDING AND SUBTRACTING DECIMALS Adding and subtracting decimals is easy as long as you line them up correctly. Consider this example: 1.287 + 24.32. First, be sure the two numbers have the same number of spaces after the decimal. The first number has three places, but the second only has two. But, you can always add zeros to the end of a decimal number, 24.32 = 24.320. Note that the last 0 means there are 0/1000, which is correct and does not change the value. Finally, it is easiest to add these numbers if you line them up vertically, always being sure the decimal points are aligned:

$$\begin{array}{r} 1.287 \\ +\ 24.320 \\ \hline =\ 25.607 \end{array}$$

Subtraction is done the same way.

PICTURE THIS

Assume that you have worked three extra jobs for pocket cash this week. From them, you earned $33.70, $45.28, and $21.02. How much extra money did you earn? ______________

With this money you buy a pizza for $8.99, soft drinks for $1.49, gas for $20.00, and a new CD for $19.95. How much money do you have left? ____________

Congratulations, you just added and subtracted decimals, as you do on a regular basis in daily life! You should see that you earned $100 and have $49.57 left.

When adding or subtracting decimals, always align the decimal point in the two numbers before doing the operation. ■

MULTIPLYING AND DIVIDING DECIMALS Multiplication and division of decimals is a bit trickier because you must keep track of how many decimal places you should have at the end. Let's try an easy one: 0.5 × 0.3. First, multiply the numbers as if they are whole numbers: 5 × 3 = 15. Now, add the number of decimal places you started with. Both numbers you multiplied originally had one decimal place, so that adds up to two. Realize that your answer of 15 is 15.0, so you know where the decimal begins. Now you have to move the decimal. The numbers you started with had a total of two digits after the decimal, so you must move the decimal point left by two places, giving you 0.15. Here is a way to double check that. If the original numbers were fractions, they would be 3/10 and 5/10. Recall how to multiply fractions—you multiply the numerators, then multiply the denominators:

$$3/10 \times 5/10 = 15/100 = 0.15$$

What if the numbers had been 0.03 × 0.5? Although you still get 15, now you need to move the decimal three places to the left, but there are only two. You simply add zeros to the left until you have the correct number of decimal places, in this case giving you 0.015.

Division with decimals is just like ordinary division, except we keep going until we either finish or reach a predetermined stopping point. You usually do not go beyond the number of decimal points your original numbers contain, so if they had a total of two, you would likely stop at two and round off beyond that. Let's look at an example: 2.8 ÷ 7. Put the decimal point in the answer line exactly above its position in the dividend (2.8), then simply do the division:

$$\begin{array}{r} 0.4 \\ 7\overline{)\,2.8} \\ -2\,8 \\ \hline 0 \end{array}$$

How do you divide when both numbers are decimals? Let's divide 2.1 by 0.7. All you have to do is move the decimal of the divisor until you have a whole number, then move the decimal of the dividend by the same number of spaces in the same direction. If you try 1.68 (dividend) ÷ 0.3 (divisor), you move the decimal in 0.3 one spot to the right to get 3, then you must also move the decimal in 1.68 one spot to the right, getting 16.8, so the problem becomes 16.8 ÷ 3. Do long division to get the result:

$$\begin{array}{r} 5.6 \\ 3\overline{)\,16.8} \\ -15 \\ \hline 18 \\ -18 \\ \hline 0 \end{array}$$

You can always multiply back to double check your result:

$$0.3 \times 5.6 = 1.68$$

Moving the decimal may seem confusing, but there are some easy shortcuts to remember.

- Moving the decimal to the *right* one space is the same as multiplying by 10; two spaces multiplies by 100; and so on, so the numbers get *bigger*.
- Moving the decimal to the *left* means you are dividing by 10 for each space moved, and the number always gets *smaller*.

Moving the decimal to the right is the same as multiplying by 10 for each space moved. Moving it to the left is the same as dividing by 10 for each space moved. ■

QUICK CHECK

Solve these problems:

1. 1.27 + 3.6 = __________
2. 14.87 − 3.2 = __________
3. 2.4 × 1.2 = __________
4. 8.4 ÷ 0.2 = __________

Answers: 1. 1.27 + 3.60 = 4.87. 2. 14.87 − 3.20 = 11.67. 3. 2.4 × 1.2 = 2.88. 4. Move the decimal in the divisor and dividend both one space to the right, so it becomes 84 ÷ 2 = 42.

PERCENTS

As you learn biology, you will encounter many values that are stated as percents. Percents are based on 100, where 100% is the total. For that reason, when working with percents, always be sure they add up to 100% and no more than that.

Percents are easy to work with. They are essentially fractions expressed as hundredths. For example, 25% is the same as 25/100, which can be further reduced to 1/4. And because fractions can be expressed as decimals, so can percents. You simply put the decimal two places to the left of the percent. So, 25% becomes 0.25 and 7% becomes 0.07. Percents, decimals, and fractions are all interchangeable, but when doing math operations, percents should be converted into either decimals or fractions. You cannot do math operations with mixed expressions—they must all be whole numbers, or fractions, or decimals. **Table 2.1** explains the relationship between these expressions.

TABLE 2.1 The relationship between percents, decimals, and fractions.

Percent	Decimal	Fraction in hundredths	Reduced fraction
10%	0.10	10/100	1/10
25%	0.25	25/100	1/4
40%	0.40	40/100	2/5
50%	0.50	50/100	1/2
75%	0.75	75/100	3/4
100%	1.00	100/100	1

TIME TO TRY

Can you supply the missing information in this table?

Percent	Decimal	Fraction
36%	______	______
______	0.42	______
______	______	80/100 = 4/5

Answers: 36% = 0.36 = 36/100 = 9/25; 42% = 0.42 = 42/100 = 21/50; 80% = 0.8 = 80/100 = 4/5.

Can You Feel the Power? **Understanding Exponents**

We briefly discussed exponents earlier, when we stated that a number written with an exponent is basically a multiplication problem: $2^3 = 2 \times 2 \times 2 = 8$. In science, very large and very small numbers are often written in a special format that uses exponents based on powers of 10. This format is called **scientific notation.** Let's consider the number 200 to see how this format is used. To write a number in scientific notation, first place the decimal immediately after the first digit, and then drop the zeroes. This number is called the **coefficient**. In this example, the coefficient is 2. Next, count how many spaces you moved the decimal—two places to the left. Each of those spaces represents a power of 10, so two places means $10 \times 10 = 100$. In scientific notation, we would write 200 as 2×10^2. As you can see, this is $2 \times 10 \times 10 = 200$. Remember, the first number in scientific notation must be greater than 1 but less than 10. If a number is less than 1, the exponent is a negative power of 10. For example, 0.0004 would be 4×10^{-4} because the decimal was moved four spaces to the right. You will rarely need to do math operations with scientific notation, so we will skip those. You should, however, understand scientific notation so that you can understand some of the measurements you will read about—such as cell sizes, which are often measured in micrometers (10^{-6} meter). **Table 2.2** lists some common exponents.

TIME TO TRY

Express the two numbers below in scientific notation.

24,000,000 = ___________

0.003 = ___________

If you did this correctly, you got 2.4×10^7 and 3×10^{-3}.

TABLE 2.2 The values of some common exponents used in scientific notation.

Exponent	Value	Term
10^9	1,000,000,000	billions
10^6	1,000,000	millions
10^3	1000	thousands
10^2	100	hundreds
10^1	10	tens
10^0	1	ones
10^{-1}	1/10	tenths
10^{-2}	1/100	hundredths
10^{-3}	1/1000	thousandths
10^{-6}	1/1,000,000	millionths
10^{-9}	1/1,000,000,000	billionths

My Cell's Bigger than Your Cell! **Ratios and Proportions**

- Sodium and potassium move across cell membranes in a 3:2 relationship.
- In the United States, the ratio of males to females at birth is about 105:100.
- The ratio of males to females declines steadily until, after age 85, it is only 40.7:100.

HUH? Welcome to the comparatively interesting realm of ratios. A **ratio** expresses a relationship between two or more numbers—it is a way to compare them. Ratios can be expressed using a colon between the numbers (as above), as a fraction, or by using the word "to." For example, carbohydrates contain hydrogen and oxygen in a 2 to 1 ratio, meaning there are twice as many hydrogen atoms in carbohydrates as there are oxygen atoms.

Ratios are used for comparison, and they can also be expressed as fractions. For example, a ratio of 1:2 means the same as 1/2. Look at that carefully, though. Let's say the ratio of men to women in your biology class is 1:2. We're not saying that half of the class are men, we are saying there are half as many men as women.

If ratios can be expressed as fractions, they can also be expressed as decimals and percents. Because they can be written as fractions, they can also be reduced like fractions. For example, a ratio of 4:6 is the same as 4/6, which is the same as 2/3. When working with ratios, it is critical to write them in the correct order. If a biology class has 10 males and 20 females, the ratio of males to females is 10:20. If we write it as 20:10, it means there are twice as many males as females, which is not true.

TIME TO TRY

Empty your pocket or purse of change. Separate the coins by denomination. Count all of the coins in each category. Now express those numbers in a ratio: ____________

Why is it important to indicate the order in which you are listing the coins? ____

____

If you did this correctly, you should have indicated the order of the coins, because without that reference, we have no idea what number corresponds with which coin. Perhaps you had 5 pennies, 4 nickels, 3 dimes, and 2 quarters. If you wrote your ratio in that order, it would be 5:4:3:2.

We also use ratios to discuss quantities in a certain amount. For example, there are about 280 million hemoglobin molecules in each red blood cell. That can be expressed as 280 million/cell, which looks more like a fraction but is read as 280 million per cell; it is really saying the ratio is 280 million to 1. Rates are also a special type of ratio. A red blood cell travels about 700 miles in its 120-day lifespan, giving a ratio of 700 miles/120 days.

Proportions are statements of equal ratios. A simple example would be to say $1/2 = 4/8$. In science, we often use proportions to solve problems. To see how, examine a generic version:

$$\frac{a}{b} = \frac{c}{d}$$

Because these two ratios are equal, their cross-products are also equal, due to some basic laws of math. This means that the product of multiplying the first numerator (a) by the second denominator (d) equals the product of multiplying the second numerator (c) by the first denominator (b):

$$a \times d = c \times b$$

Let's say we want to know how many times the heart beats in an hour. Assume that the heart beats on average 80 beats per minute. We know there are 60 minutes per hour. So, we can set up the proportion, filling in the information we know and using "x" to represent the value we are trying to determine:

$$\frac{80 \text{ beats}}{1 \text{ minute}} = \frac{x \text{ beats}}{60 \text{ minutes}}$$

We know we can **cross-multiply (Figure 2.4)**. When we do that, we get $4800 = x$, so there are 4800 beats per 60 minutes, or per hour. In fact, there is a simpler way to write this problem, which is a shorter version of cross-multiplying. We know there are 80 beats per minute, and 60 minutes per hour, so we can calculate the beats per hour as follows:

$$\frac{80 \text{ beats}}{\text{minute}} \times \frac{60 \text{ minutes}}{\text{hour}} = \frac{4800 \text{ beats } \cancel{\text{minutes}}}{\cancel{\text{minute}} \text{ hour}} = \frac{4800 \text{ beats}}{\text{hour}}$$

Notice that the units are shown, and in the next to last step, minutes appear on both the top and bottom. That means they cancel each other out, so we are left with beats per hour, which is the correct unit. Using the units can be an easy way to ensure that you have set up the problem correctly. This is another example of taking the time to think through the problem before you start—the units should make sense when you are done.

1. Set up the proportion with what is known, using "x" to represent the information you are seeking.

$$\frac{80 \text{ beats}}{1 \text{ minute}} = \frac{x \text{ beats}}{60 \text{ minutes}}$$

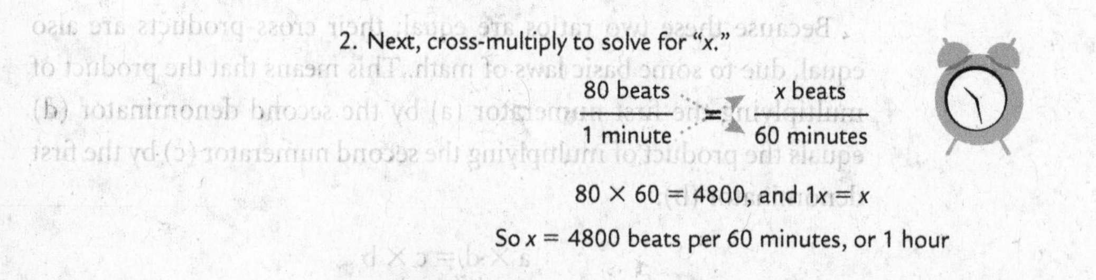
2. Next, cross-multiply to solve for "x."
80 beats / 1 minute = x beats / 60 minutes
80 × 60 = 4800, and 1x = x
So x = 4800 beats per 60 minutes, or 1 hour

FIGURE 2.4 Using proportions to determine how many times the heart beats in an hour.

✔ **QUICK CHECK**

An average person takes about 12 breaths per minute. How many breaths do they take in an hour? ________

Answer: Set up the proportion: 12 breaths/minute = x breaths/60 minutes.
Cross-multiply: $12 \times 60 = 720 = x$, so $x = 720$ breaths per hour.

Who Ever Heard of a Centimeter Worm? **The Metric System**

In the United States, we all grew up learning there are 12 inches to a foot, 3 feet to a yard, and 100 yards to a football field, and we measure driving distances in miles, which contain 5280 feet. In the kitchen, we use cups, half cups, thirds of cups, quarter cups, tablespoons, teaspoons, eighths of teaspoons, pints, quarts, and gallons. There are so many units in our system it is amazing we can keep them straight. But of course there is a simpler way to measure.

It is called the **metric system**, or **System Internationale (SI)**. It is universally used in science and by almost every country in the world except the United States. You have undoubtedly had brushes with learning the metric system, and you may have found it difficult. The problem is not with the metric system, which is delightfully simple. Instead, the problem is with our complicated U.S. (also called English) system and the need to convert between the two systems. This requires—you guessed it—math.

WHY SHOULD I CARE?

Science uses metric measurement almost exclusively, so you will need a basic understanding of metric units for all your future coursework. In addition, almost everyone on our planet—except the United States—uses the metric system.

The metric system is amazingly simple because it is all based on the number 10, which means decimals are easier to use than they are in the U.S. system. For our purposes, we'll learn four main units used in science: those that deal with length or distance, mass, volume, and temperature. Each of these has a standard or base unit:

- The basic unit of length (or distance) is the **meter (m)**.
- The basic unit of mass is, technically, the **kilogram (kg)**, but many sources use the **gram (g)** as the base unit instead.
- The basic unit of volume is the **liter (L)**.
- The basic unit of temperature is the **degree Celsius (°C)**.

These are the base units, but more convenient units are derived from these. For example, a meter is just a bit longer than 3 feet (39.34 inches), so it is not a convenient unit for measuring the size of, say, your finger or a cell. Smaller units of the meter, based on the powers of 10, are used instead. These units are named by adding the appropriate prefix to the term *meter* (Table 2.3). **Centi-** means 1/100, and there are 2.54 centimeters (cm) in an inch, so centimeters work well for measuring fingers. Cells are microscopic, so they are best measured in even smaller units, such as micrometers—one micrometer = 1 millionth of a meter. Driving between cities, you can best measure the long distances in kilometers, each of which equals 1000 meters. Again, all metric units are based on 10. Think about that—you first learn to count from 1 to 10, then you can count by tens to 100, then by hundreds to 1000, and so on. It is an easy system.

Table 2.3 provides many of the prefixes and their base-10 equivalent. In biology, you will use some units more often than others. For length or distance, which is a straight linear measurement, you will mostly work in meters, centimeters, millimeters (1/1000 m), and micrometers. For mass, which is the actual physical amount of something, you will most often refer to kilograms (1000 grams), grams, and milligrams (1/1000 g). For volume, which refers to the amount of space something occupies, the most common units will be liters and milliliters.

TABLE 2.3 Metric system prefixes.

Prefix	Symbol	Decimal equivalent (multiple)	Exponential equivalent (scientific notation)
Pico-	p	0.000000000001	10^{-12}
Nano-	n	0.000000001	10^{-9}
Micro-	μ	0.000001	10^{-6}
Milli-	m	0.001	10^{-3}
Centi-	c	0.01	10^{-2}
Deci-	d	0.1	10^{-1}
no prefix		1.0	10^{0}
Deka-	D	10.	10^{1}
Hecto-	H	100.	10^{2}
Kilo-	k	1000.	10^{3}
Mega-	M	1,000,000.	10^{6}
Giga-	G	1,000,000,000.	10^{9}

The Celsius temperature scale does not use prefixes. Instead, it has a single unit: degrees Celsius. However, the scale is different than the Fahrenheit scale with which you are familiar. Water freezes at 0°C and boils at 100°C (**Figure 2.5**).

At what Fahrenheit temperature does water freeze? ______________ Boil? ________ In the Celsius scale, normal body temperature (98.6°F) is 37°C. To convert between the two temperature scales, there are two specific equations—one to convert from degrees Celsius to degrees Fahrenheit, and another to convert in the opposite direction. Each of these equations is listed below, first in its original form, which includes a fraction, and then with the fraction converted to a decimal. You will likely use a calculator to do any conversions, and it will be easier to multiply using the decimal.

$$
\begin{aligned}
°\text{Celsius} &= (°\text{Fahrenheit} - 32) \times 5/9 \\
&= (°\text{Fahrenheit} - 32) \times 0.556
\end{aligned}
$$

$$
\begin{aligned}
°\text{Fahrenheit} &= °\text{Celsius} \times 9/5 + 32 \\
&= °\text{Celsius} \times 1.8 + 32
\end{aligned}
$$

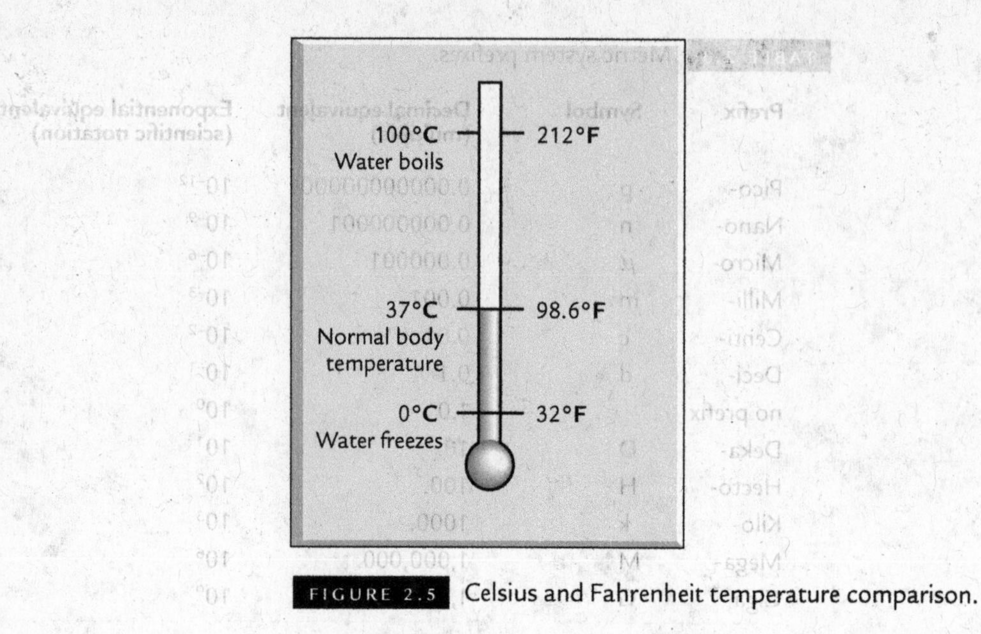

FIGURE 2.5 Celsius and Fahrenheit temperature comparison.

Let's try one of each type of conversion. To convert 37°C to °F, we use the second equation:

$$
\begin{aligned}
°F &= 37°C \times 1.8 + 32 \\
&= 66.6 + 32 \\
&= 98.6°F
\end{aligned}
$$

Now let's convert 75°F to °C. We use the first equation as follows:

$$
\begin{aligned}
°C &= (75°F - 32) \times 0.556 \\
&= 43 \times 0.556 \\
&= 23.9°C
\end{aligned}
$$

Most of the work you will do, though, will be converting units for length, mass, and volume, so let's move on.

TIME TO TRY

You will use the metric system more easily if you have some idea of the size of the base units. Most packaged items manufactured in the United States list both metric and U.S. units. Most rulers and measuring tapes have both metric and U.S. units. Explore your home and see if you can determine the following:

- Which is longer, a meter or a yard? ___________
- Which is heavier, a gram or a pound? ___________

Answer: A meter is longer than a yard by a few inches. A pound contains just over 450 grams.

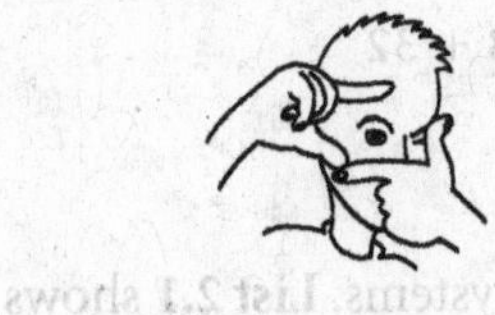

PICTURE THIS

If you drink carbonated soft drinks, you are likely quite familiar with their standard large plastic bottles. What is their volume in metric units? ______ liters

Knowing this, envision the bottle only half full. Is that amount more or less than a gallon of milk? ______

Answer: Half of a 2-L soda bottle is definitely less than a gallon of milk, so a liter is smaller than a gallon.

Know the paper clip! A standard small paper clip has a mass of about 1 g (it's very light). A standard large paper clip has a side-to-side width of about 1 cm, and the wire from which it is made has a diameter of about 1 mm.

You will become more familiar with the metric system as you use it. You may occasionally need to convert from U.S. units to metric units, although this is done more as an exercise than out of need—in class almost all measuring and discussion will use metric units. Still, it is

LIST 2.1 Some basic conversions.

SPECIAL RELATIONSHIPS

1 milliliter (mL) = 1 cubic centimeter (cc), a unit often used in administering liquid medications.

The mass of 1 milliliter of water = about 1 gram.

The mass of 1 liter of water = about 1 kilogram.

APPROXIMATE CONVERSION FACTORS

Multiply **inches** × 2.54 cm/inch to get **centimeters**.

Multiply **feet** × 0.305 m/foot to get **meters**.

Multiply **miles** × 1.6 km/mile to get **kilometers**.

Divide **pounds** by 2.2 pounds/kg to get **kilograms**.

Multiply **gallons** × 3.8 L/gallon to get **liters**.

°Celsius: (°Fahrenheit − 32) × 5/9 = (°Fahrenheit − 32) × 0.556

°Fahrenheit: °Celsius × 9/5 + 32 = °Celsius × 1.8 + 32

useful to know how to convert between the two systems. **List 2.1** shows some of the basic conversion factors.

Let's try some conversions. First, let's do the easy stuff: converting between metric units. Remember that the difference between the units will always be some multiple of 10. Let's convert 13 meters into centimeters.

$$13 \text{ m} = \_\_\_\_\_\_\_\_ \text{ cm}$$

A centimeter is 1/100 of a meter, so there are 100 centimeters per meter. Thus:

$$13 \text{ m} \times 100 \text{ cm/m} = 1300 \text{ cm}$$

Now we will convert 27 millimeters into centimeters, but let's try another method. All we really have to do to convert between different metric units is move the decimal, but by how many spaces and in which direction? How many spaces you move is determined by the difference in the power of 10. We know that millimeters are thousandths of a meter, and centimeters are hundredths of a meter.

$$\text{millimeters} = 10^{-3} \qquad \text{centimeters} = 10^{-2}$$

So, if we look at the exponents, they are different by one. We will move the decimal in our number (27) by one spot. But in which direction? When converting from smaller to larger units, the decimal moves left. When converting from larger to smaller units, the decimal moves right. Back to our example: Converting 27 mm to cm gives us 2.7 cm.

When converting within metric units:

1. Put the units in scientific notation and subtract the smaller exponent from the larger one. The difference is how many spaces the decimal will move in your coefficient.
2. If you are converting from small units to larger ones, the number gets smaller, so the decimal moves to the left. If you are converting from larger units to smaller ones, the number gets bigger, so the decimal moves to the right. ■

TIME TO TRY

Now that you see the simple secret to this process, complete the following conversions:

5 kg = ________ g   8 mL = ________ L   6 cm = ________ m

If you did these conversions correctly, you should see that 5 kg = 5000 g; 8 mL = 0.008 L; 6 cm = 0.06 m. See, the metric system is easy!

Now let's convert from metric to U.S. units, and vice versa. To do these you need the correct conversion factor (see List 2.1). Let's convert 18 inches into centimeters. From the list we see that there are 2.54 cm per inch, so we merely multiply 18 inches × 2.54 cm/inch. The answer is 45.72 cm.

Let's convert 30 cm into inches. We can try a proportion to solve this.

$$\frac{2.54 \text{ cm}}{1 \text{ inch}} = \frac{30 \text{ cm}}{x \text{ inches}}$$

Cross-multiplying gives us 2.54 x = 30, so to get x we divide both sides by 2.54: 30 ÷ 2.54 = 11.81 inches. You now have all the tools you need to do conversions between the two systems. The easiest solution, however, is to only work in metrics, like the rest of the world!

✔ **QUICK CHECK**

Complete these conversions:

1. 30 miles = ________ km
2. 8 L = ________ mL
3. 110 pounds = ________ kg

Answers: 1. 30 mi × 1.6 km/mi = 48 km. 2. 8 L × 1000 mL/L = 8000 mL.
3. 110 pounds ÷ 2.2 pounds/kg = 50 kg.

How Do You Measure Up? **Basic Measurement**

Now that you understand the basic units of measurement, you need to know how to measure. A common error in scientific experimentation is called human error, which includes math mistakes (which you won't make now!) and something as simple as not measuring correctly. When you're baking brownies, adding extra sugar and chocolate may be a good thing, but that won't work in science. Measurements must be done precisely and with appropriate equipment.

MEASURING LENGTH

Length is usually measured with a meter stick or a ruler. Grab one—surely you can locate one somewhere.

Examine the scale. There are likely two scales—inches and metric. On a meter stick, the scales are often on opposite sides of the stick. Look at the metric scale. If it is a meter stick or metric tape measure, the small numbered units are usually centimeters. Confirm that there are 100 of these in a meter. Compare the size of a centimeter to an inch by placing your fingers on each side of a centimeter, then maintaining that space as you move to the inch scale. Note that an inch is a bit over 2.5 times bigger—2.54 to be exact (**Figure 2.6a**).

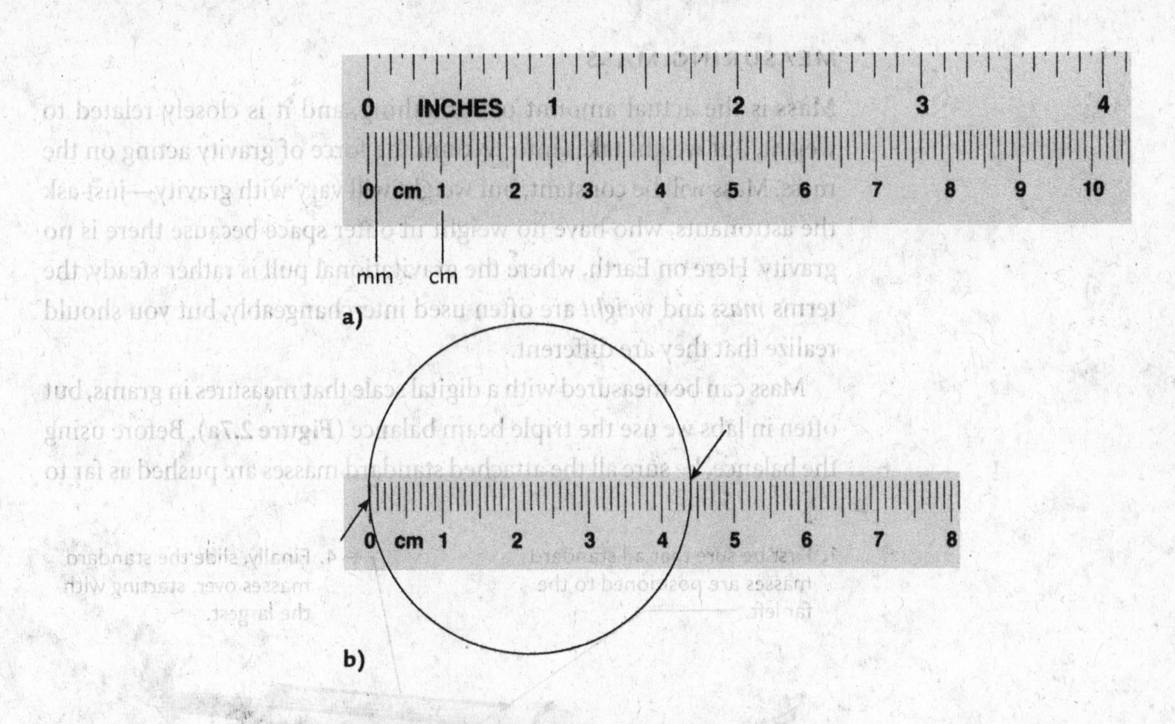

FIGURE 2.6 Using a metric ruler. **a)** This shows the comparison of the metric and inch scales. Millimeters and centimeters are labeled. **b)** Always start measuring at the very edge of the object, positioning the ruler carefully so that the scale runs precisely through the dimension being measured.

Now examine the space between 0 and 1 cm. Count the spaces. How many are there? ______ You should have counted 10 spaces. These tiny units are millimeters, each equal to 1/1000 meter. That makes sense—each millimeter is 1/10 of a centimeter, which is 1/100 of a meter, and 1/10 of 1/100 is, indeed, 1/1000.

When measuring with a ruler, be sure to align the zero line exactly at the edge of your object and measure exactly to the far edge. Do not round off. If you are measuring the diameter of a circle, be sure your ruler is positioned across the widest part of the circle (**Figure 2.6b**).

When measuring length, be sure you use the appropriate scale.

MEASURING MASS

Mass is the actual amount of something, and it is closely related to weight, but weight takes into account the force of gravity acting on the mass. Mass will be constant, but weight will vary with gravity—just ask the astronauts, who have no weight in outer space because there is no gravity. Here on Earth, where the gravitational pull is rather steady, the terms *mass* and *weight* are often used interchangeably, but you should realize that they are different.

Mass can be measured with a digital scale that measures in grams, but often in labs we use the triple beam balance (**Figure 2.7a**). Before using the balance, be sure all the attached standard masses are pushed as far to

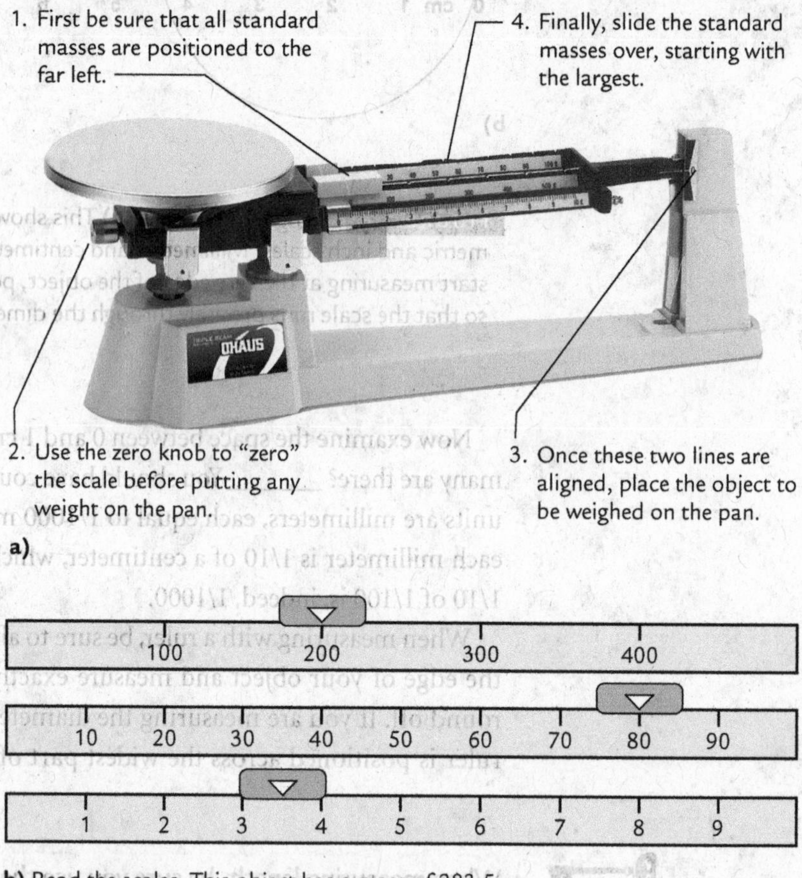
1. First be sure that all standard masses are positioned to the far left.
4. Finally, slide the standard masses over, starting with the largest.
OHAUS
2. Use the zero knob to "zero" the scale before putting any weight on the pan.
3. Once these two lines are aligned, place the object to be weighed on the pan.
100 200 300 400
10 20 30 40 50 60 70 80 90
1 2 3 4 5 6 7 8 9

a)

b) Read the scales. This object has a mass of 283.5 g.

FIGURE 2.7 Using the triple beam balance.

the left as possible. Next, you must "zero" the balance before adding anything to the pan. Notice how the two lines on the far right line up. One line is on the arm and moves with it; the other is on the end piece of the scale. If they are not perfectly aligned, slowly turn the zero knob located on the left of the scale, usually under the pan. Rotate the knob in either direction until the two lines are aligned. The arm will move up and down a bit—wait until it has stopped and the lines are aligned. The scale is now zeroed. NOW you are ready to measure the mass of your object.

There are three beams on the arm of the balance, each suspending a different standard mass. The largest mass is 100 g. Each spot that you move that mass to the right equals 100 g. Another beam has a 10-g mass, and the front beam has a 1-g mass. Starting with the 100-g mass, slide it across until it causes the arm to swing too far to the right. That was too heavy, so back it up to the left by one spot. Note the number—that is how many hundred grams are in your object. Next, slide the 10-g mass until it is too much, back up one notch, then note how many 10's of grams there are. Finally, slide the 1-g mass carefully until the two lines on the right side again align, as they did at the beginning. At that point, the scale is balanced. Read all of the whole numbers on the scale; each line beyond the last whole number is 1/10 g. If the 100-g mass is at 200, the 10-g mass is at 80, and the 1-g mass is halfway between the 3 and 4, what is the mass of the object?

(It would be 283.5 g, as shown in **Figure 2.7b**.)

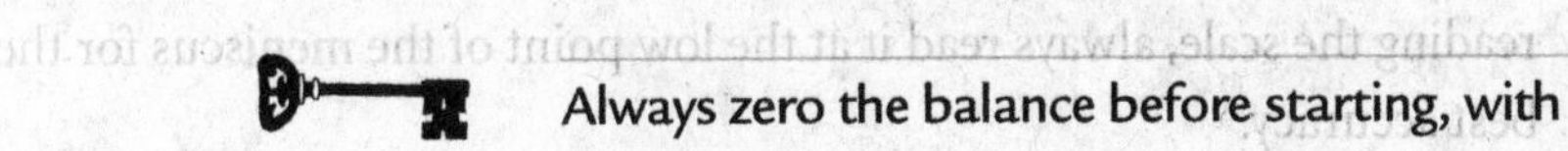

Always zero the balance before starting, with all standard masses at the far left. ■

MEASURING VOLUME

Volume refers to the amount of space a substance occupies. In biology labs, you will most often measure liquid volumes by using beakers or graduated cylinders. Always use the smallest container in which the substance will fit—the smaller it is, the more accurate your reading will be. Always read the scale, usually in milliliters, at eye level for accuracy. You should also know how to read the meniscus (**Figure 2.8**)—this is especially critical in a graduated cylinder. Liquid in a container tends to

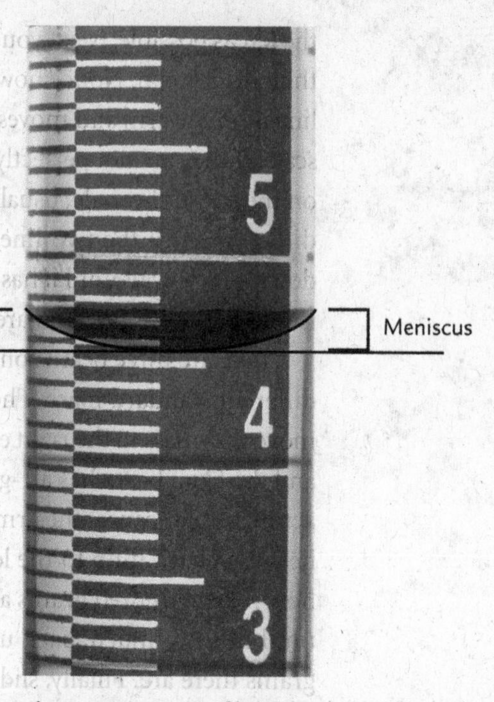

FIGURE 2.8 Read measurements of liquid volumes from the bottom of the meniscus. By doing so, you see that this graduated cylinder contains 4.55 mL of liquid, not the 4.75 indicated by the top of the meniscus.

climb slightly up the sides, so the center is lower than the edges, where the liquid contacts the container. This dip is called the **meniscus**. When reading the scale, always read it at the low point of the meniscus for the best accuracy.

TIME TO TRY

Find the narrowest clear container you can and fill it halfway with water. Look at it at eye level. Use a ruler on the outside of the container to measure the highest point of the water. __________ Now measure the lowest point. ________ The dip that you see is the meniscus. Whenever you measure a liquid volume, always measure at the lowest point of the meniscus.

You Ought to be in Pictures: **Tables, Graphs, and Charts**

We explored how to get numbers by measuring and how to work with them. Now we will see how these numbers and other information, collectively called data, can be depicted.

TABLES

We have already used tables in this book, and you should be familiar with them, so let's quickly review the basics. Tables come in many forms and are a convenient way to present information so it is easy to read and compare. We will use **Table 2.4** as a reference.

When viewing a table, start with the table title, in this case "Fluctuations in human body temperature throughout the day." The title usually tells you what the table contains. Next, realize that tables are carefully arranged in columns and rows. All information in a single column is related, and all information in a single row is related. Look at the top of each column—these are column heads that tell you what information each column contains. Look at the beginning of each row. These are row heads, or labels that tell you what each row contains. In our example, the column heads reveal that the first column identifies each test subject, and the other columns contain the temperatures for all test subjects at specific times of the day. The row heads tell us that all temperatures in a single row belong to a single test subject, and who it is. So, by using the column and row heads, it is easy to find out, for example, what temperature Subject C had at noon.

TABLE 2.4 Fluctuations in human body temperature throughout the day.

Subject	T °F at 4 AM	T °F at 8 AM	T °F at noon	T °F at 4 PM	T °F at 8 PM	T °F at midnight
A	96.0°F	96.3°F	96 4°F	96.8°F	97.0°F	96.8°F
B	98.0°F	98.4°F	98.6°F	98.8°F	99.0°F	98.4°F
C	97.6°F	98.0°F	98.4°F	98.5°F	98.9°F	98.6°F
D	97.2°F	97.3°F	97.5°F	97.6°F	98.0°F	97.6°F

TIME TO TRY

On a separate piece of paper, using sentences and paragraphs, write out all the information that is included in Table 2.4. Which version is easier to read? In which version can you more easily determine Subject D's temperature at 8 PM? ______________

GRAPHS

Graphs present a more pictorial view of data. Numerical data that can be organized in a table (**Figure 2.9a**) can usually also be presented in a graph. The main advantage is that the graph allows you to spot trends and relationships almost instantly. There are various types of graphs. Let's look at three of them: line graphs, bar graphs, and pie charts.

LINE GRAPHS Look at **Figure 2.9b**. This is a line graph, and they usually are laid out in a grid. The horizontal axis at the bottom is the

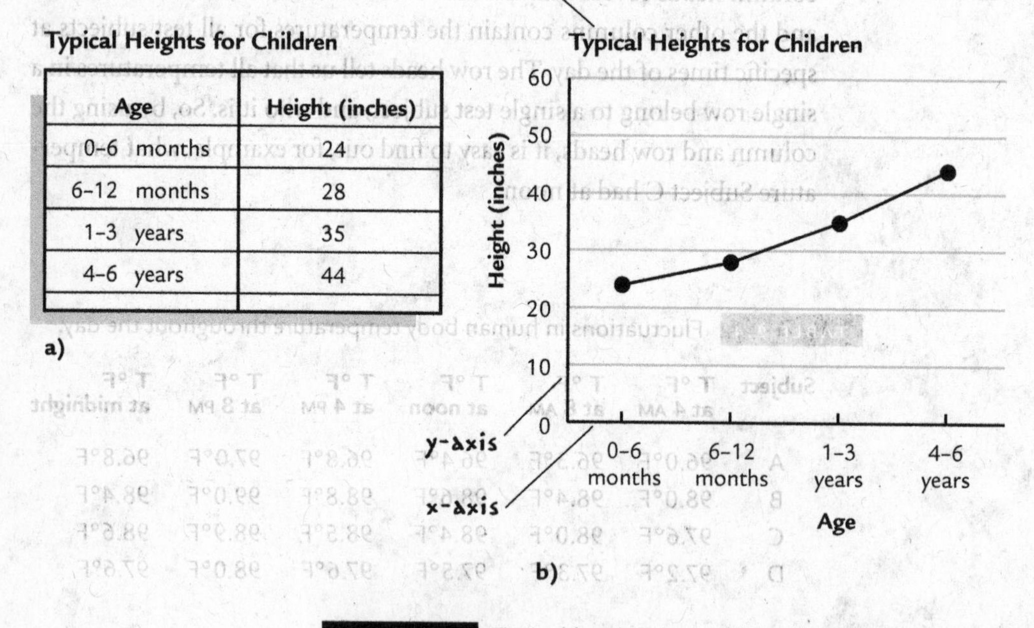

Typical Heights for Children

Age	Height (inches)
0–6 months	24
6–12 months	28
1–3 years	35
4–6 years	44

FIGURE 2.9 Typical heights for children through age 6 years. **a)** Table format. **b)** Basic line graph.

x-axis. It often, but not always, marks the progression of time. The vertical axis on the left is the ***y*-axis** and it typically reflects some increasing value. Where these lines meet on the lower left of the graph marks the 0 position, so units go up as you move away from that point. Each axis should be labeled and should include units.

In our example, the *x*-axis tells us who the data are about—the age group of children for whom the listed height is typical. The *y*-axis gives height in inches. For each age group, the height (data point) is placed above the age it represents. The data points may be left unconnected, they may be directly connected (as they are here), or a line of "best fit" may be drawn that passes between the points so they are evenly distributed on each side of it. In our example, because this depicts growth with time, the line allows us to see at once the heights for each age group and to quickly comprehend the trend of a gradual increase in height with age.

When drawing a line graph, don't forget to label the axes and to include units. The most common mistake made when graphing data is to not use an appropriate scale. Be sure to size the units to maximize the space the graph fills. You don't want the graph to be cramped into one corner, making it hard to read. Spread it out both vertically and horizontally.

BAR GRAPHS **Figure 2.10** is a bar graph showing the amount of water in the human body. Looking at the axes, you see that the *x*-axis has three separate categories: total water, intracellular fluid (the water located inside body cells), and extracellular fluid (the water not contained in cells). Each of these categories has two bars—one for males and one for females. The *y*-axis tells what percent of the total body weight the water represents. This graph is drawn so that the male and female data are directly compared by being positioned side-by-side, yet easily distinguished by use of different shading. The shading is explained on the lower left in a feature called the **key**. This bar graph has a 3-D effect that does not change its meaning at all—it just makes it slightly more interesting. Bar graphs may be drawn vertically or horizontally. Because each bar is so distinct, these graphs are good for comparing specific bits of data, whereas line graphs may be better at showing an overall trend.

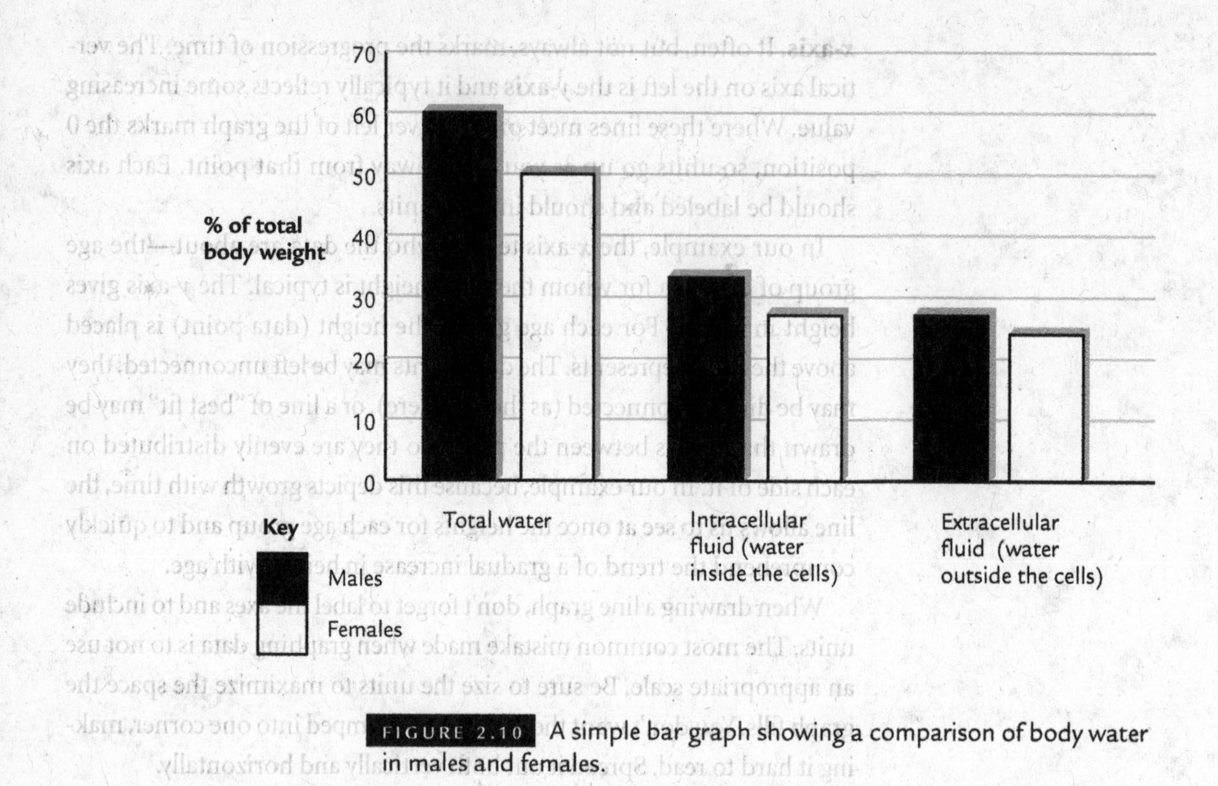

FIGURE 2.10 A simple bar graph showing a comparison of body water in males and females.

PIE CHARTS **Figure 2.11** is a pie chart, a type of graph that shows parts of a whole. This one shows the main molecules that make up the human body and tells the percent of the body made of each type of molecule. When looking at this, you immediately see that all the parts add up to the whole "pie," which is 100%, so these charts are very effective when showing percents. We all have a visual concept of a whole pie and a slice of pie, so even before looking at the numbers, we instantly see the differences in distribution. You know right away from this pie chart that one type of molecule makes up well over half of the human body (see Chapter 5, Chemistry, for a refresher course on molecules and related topics).

Pie charts have no axes to label and the space within them is limited, so a key is often used. Labels for each "slice" may be written within the pie or placed to the outside, as in our example. Using different shading or colors for the slices makes these graphs even more readable. Because there are no axes to provide information, the graph title and key are very important.

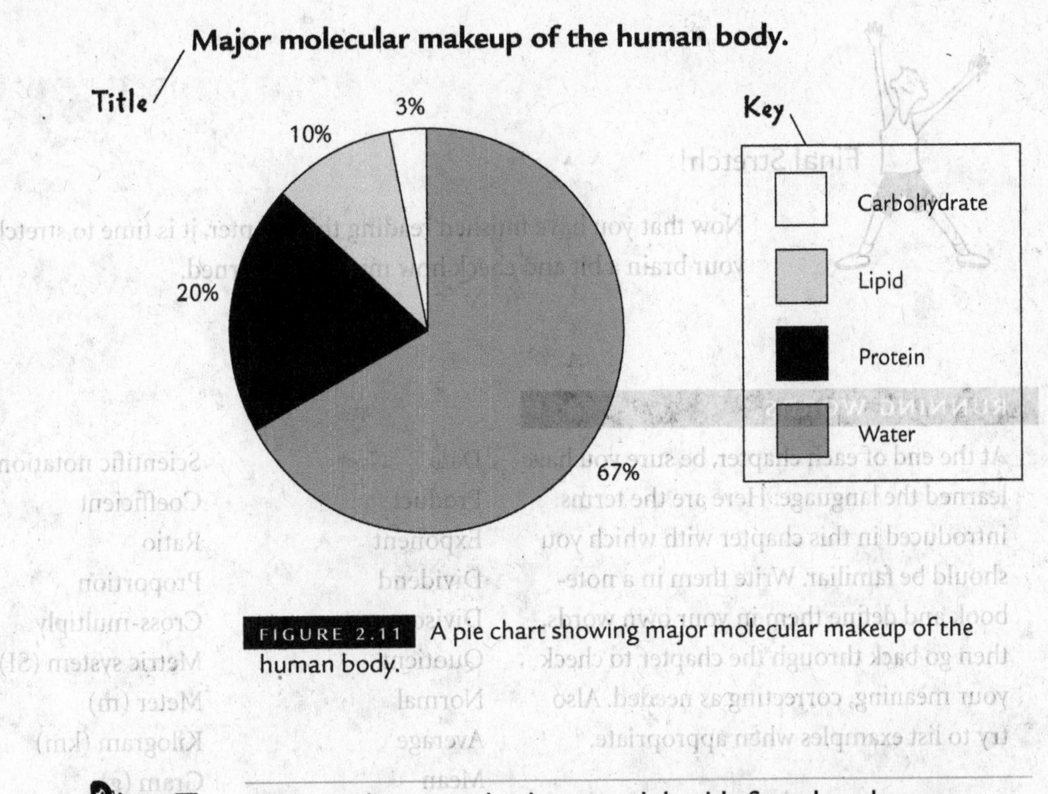

FIGURE 2.11 A pie chart showing major molecular makeup of the human body.

When reading a graph, always read the title first, then the axes, key, and labels. Finally, just let your eyes take in the relationships depicted. ■

✔ **QUICK CHECK**

1. In Figure 2.9, which age group is shortest? ____________ What is the typical height for children aged 1 to 3 years? ____________
2. In Figure 2.10, what percent of the total body weight is water in an average female? ____________
3. In Figure 2.11, which molecule makes up most of the human body? ____________

Answers: 1. The shortest group is aged 0–6 months, and children aged 1–3 are typically 35 inches tall. 2. 50% 3. water

Final Stretch!

Now that you have finished reading this chapter, it is time to stretch your brain a bit and check how much you learned.

RUNNING WORDS

At the end of each chapter, be sure you have learned the language. Here are the terms introduced in this chapter with which you should be familiar. Write them in a notebook and define them in your own words, then go back through the chapter to check your meaning, correcting as needed. Also try to list examples when appropriate.

Data
Product
Exponent
Dividend
Divisor
Quotient
Normal
Average
Mean
Numerator
Denominator
Equivalent fraction
Reduce
Greatest common factor (GCF)
Common denominator
Least common multiple (LCM)
Repeating decimal
Scientific notation
Coefficient
Ratio
Proportion
Cross-multiply
Metric system (SI)
Meter (m)
Kilogram (km)
Gram (g)
Liter (L)
Degree Celsius (°C)
Centi-
Milli-
Kilo-
Mass
Volume
Meniscus
x-axis
y-axis
Key

WHAT DID YOU LEARN?

Try these exercises from memory first, then go back and check your answers, looking up any items that you want to review. Answers to these questions are at the end of the book.

PART A: SOLVE THESE PROBLEMS.

1. $(4 \div 2) + 6 - 5 \times 2^3 =$ ________
2. $2 \times 10^4 =$ ________
3. 27/36 reduced is ________
4. $3/8 \times 2/3 =$ ________
5. $5/6 - 7/12 =$ ________
6. $0.5 \times 0.4 =$ ________
7. If the respiratory rate is 12 breaths per minute, how many breaths are taken in 1 hour? ________
8. 4 meters = ________ centimeters
9. 1 inch = 2.54 cm, so 1 foot = ________ cm
10. If a man weighs 200 pounds and 60% of his body weight is water, how many pounds of water does he have?

    ____

PART B: ANSWER THESE QUESTIONS.

1. What is the mean of 8, 9, 12, 18, and 23?

   ____
2. What is the numerator in 4/5? ________
3. Express 3/10 as a decimal ________ and as a percent ________
4. From Figure 2.11, how much of the human body is made of protein?

   ____
5. What would you be measuring if you are looking at a meniscus? ________
6. In the metric system, list the base unit for each of the following:

   mass: ________

   length: ________

   volume: ________
7. What is meant by "normal" blood pressure? ________
8. At what Celsius temperature does water boil? ________
9. The amount of space something occupies is called ________.
10. How many milligrams are there in 1 gram?

    ____

WHAT DID YOU LEARN?

Try these exercises from memory first, then go back and check your answers, looking up any items that you want to review. Answers to these questions are at the end of the book.

PART A: SOLVE THESE PROBLEMS.

1. $(4 + 2) + 6 - 5 \times 2^2 =$ ______
2. $2 \times 10^4 =$ ______
3. 27/36 reduced is ______
4. $3/8 \times 2/3 =$ ______
5. $5/6 - 7/12 =$ ______
6. $0.5 \times 0.4 =$ ______
7. If the respiratory rate is 12 breaths per minute, how many breaths are taken in 1 hour? ______
8. 4 meters = ______ centimeters
9. 1 inch = 2.54 cm, so 1 foot = ______ cm
10. If a man weighs 200 pounds and 60% of his body weight is water, how many pounds of water does he have? ______

PART B: ANSWER THESE QUESTIONS

1. What is the mean of 8, 9, 12, 18, and 23? ______
2. What is the numerator in 4/5? ______
3. Express 3/10 as a decimal ______ and as a percent ______
4. From Figure 2.11, how much of the human body is made of protein? ______
5. What would you be measuring if you are looking at a meniscus? ______
6. In the metric system, list the base unit for each of the following:
   mass: ______
   length: ______
   volume: ______
7. What is meant by "normal" blood pressure? ______
8. At what Celsius temperature does water boil? ______
9. The amount of space something occupies is called ______
10. How many milligrams are there in 1 gram? ______

3 Terminology

The ABCs of Biology

When you complete this chapter, you should be able to:

- Break down biological terms to understand their meanings.
- Build biological terms from basic word roots, prefixes, and suffixes.

Your Starting Point

Answer the following questions to assess your knowledge of biological terminology.

1. Most biological terms arise from which languages? ______________________

2. What is the difference between an *abbreviation* and an *acryonym*? ______________________

3. What is meant by the term *hydrophobic*? ______________________

4. What is the difference between a *prefix* and a *suffix*? ______________________

5. What is an *eponym*? ______________________

How Do You Say, In Your **Language** . . .?

Have you ever read a computer manual only to find that you are still unsure of how to configure your personal firewall, or for that matter why you should? Has your mind gone numb as an auto mechanic explained all those expensive malfunctioning parts that were replaced to get rid of that mystery noise in your car? Have you tried to read the small print in an advertisement for a new prescription drug? At such times, it may seem as if other people speak a secret language. Indeed,

Answers: 1. Latin and Greek. 2. An abbreviation is a shortened form of a word or phrase; an acronym is a word formed from the first or key letters of each word in a full name. 3. *Hydro-* means "water," and *-phobic* means "fearing"; this term refers to substances such as oils that do not mix with water. 4. A prefix is a word part added at the beginning of a word, whereas a suffix is added at the end. 5. Eponyms are terms that include someone's name.

most professions have their own sublanguage, as do most academic disciplines. Biology is no exception.

Learning the terminology in this discipline may be a challenge, but we won't say it is difficult. *Difficult* leaves the option of ducking behind excuses. You may have heard students say, "I could do this if the words weren't so hard, but this is just too difficult for me!" Instead, you should think of the words as a challenge, because a challenge sparks the competitor in us—it makes us try harder. Although the words may seem intimidating at first, you will quickly learn tools and tricks to help you understand even the longest terms.

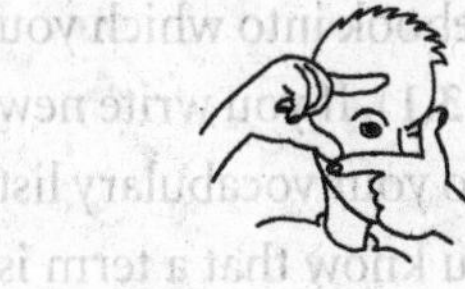

PICTURE THIS

Upon arriving at your long-anticipated vacation destination, you notice how wonderfully new and exotic everything seems–the smells in the air, the people's faces and clothing, the food sold from vendors' carts, and the sounds. You are in paradise! Then you discover that both your luggage and your wallet are missing. You approach an authority but have no idea how to speak his or her language.

1. How will you convey your plight to this person? ________________

____

2. How difficult would it be if you could speak the local language?

____

3. What could you have done before your trip to avoid this communication gap? ________________

____

Obviously if you plan to spend time in a place where another language is spoken, communication could be a challenge unless you learn some of the local language in advance. Your biology class may seem like a place where another language is spoken. Much of the terminology will be new, and most of the words do, in fact, come from languages other than English. You will learn about life by listening to lectures filled with this new language, and you will be expected to discuss your course material and write exam answers using this new language. That is why it is so important that you learn the language of biology. As you study, the first thing you should do is master—not just read, but *master*—the new words. You have to know the language before you can understand and join in the conversation.

Learning the vocabulary is the first step in learning biology. ■

We recommend that students maintain a running vocabulary list. The easiest way to do this is to keep a separate notebook into which you write all new terms as they are introduced (**Figure 3.1**). If you write new terms in your notes during lecture, transfer them to your vocabulary list and be sure to check their spellings later. Once you know that a term is spelled correctly, write the actual definition—exactly what does the word mean? You can find the actual textbook definition in your textbook or in a biological dictionary, but you should also try to explain the term in your own words.

Next, be sure you can use the word properly in a sentence. Try to add some examples that illustrate the term, if appropriate. For example, a tissue is a group of cells organized together that share a common function. Examples of tissue include bone, blood, cartilage, fat, and muscle. If you are a visual learner, try illustrating your new words if you can. If you are an auditory learner, try reading the words and their meanings out loud and consider tape recording your list. If you are a tactile learner, you might benefit from typing your list into a word-processing file, then alphabetizing it. You might also write the terms and their meanings on flashcards.

You may be surprised at the beginning of the course at the number of new terms you encounter. You can think of the words as a kind of smoke screen—the underlying principles of biology are rather simple, but you may not see that through the smoke. More than one student has jokingly commented that "It's all Greek to me." They don't realize that their joke is not far from the truth—most biological terms have either Greek

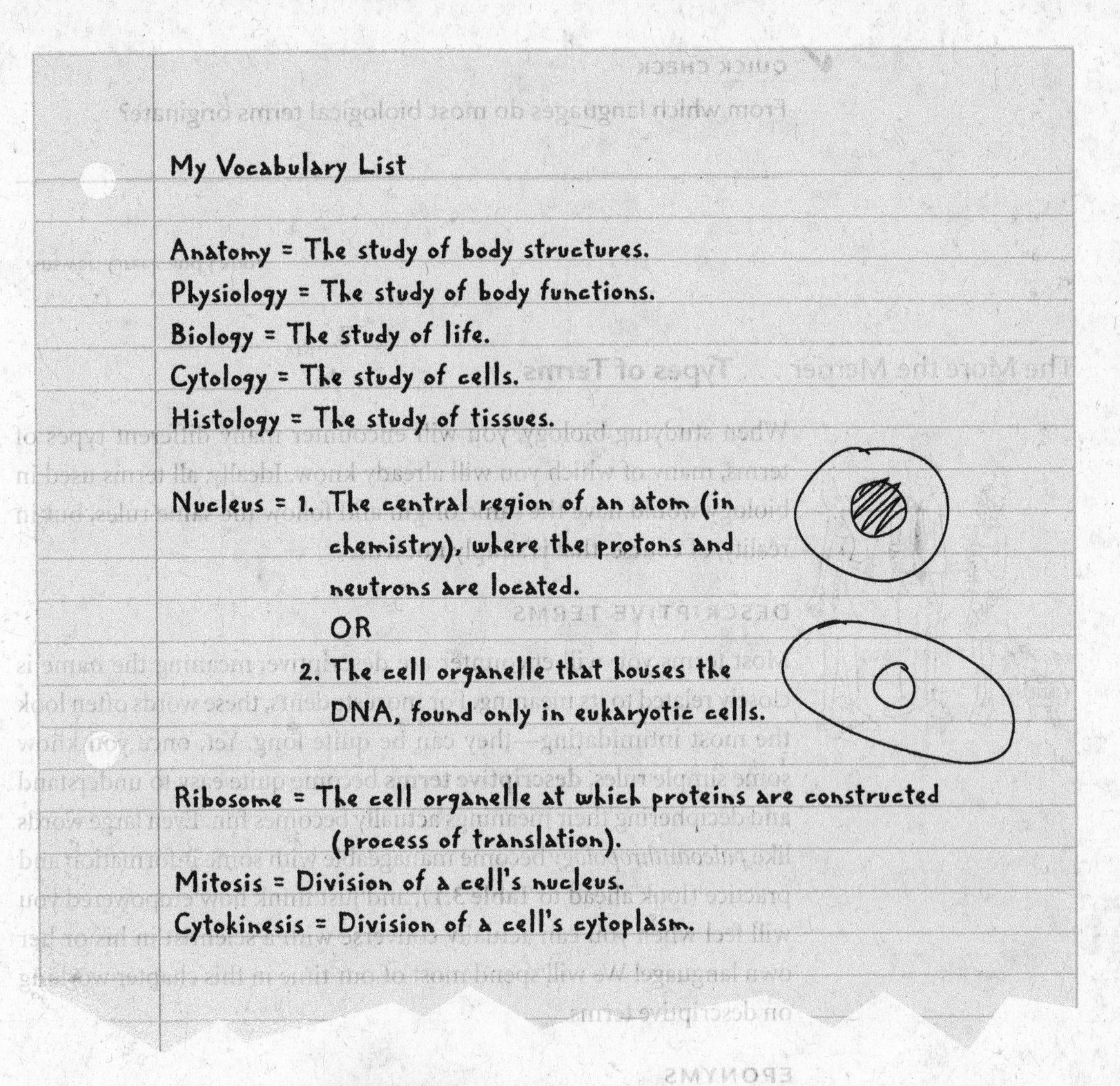
My Vocabulary List
Anatomy = The study of body structures.
Physiology = The study of body functions.
Biology = The study of life.
Cytology = The study of cells.
Histology = The study of tissues.
Nucleus = 1. The central region of an atom (in chemistry), where the protons and neutrons are located.
OR
2. The cell organelle that houses the DNA, found only in eukaryotic cells.
Ribosome = The cell organelle at which proteins are constructed (process of translation).
Mitosis = Division of a cell's nucleus.
Cytokinesis = Division of a cell's cytoplasm.

FIGURE 3.1 Keeping a separate running vocabulary list throughout the semester can help you master the language.

or Latin origins, although many terms are derived from other languages as well. So, instead of thinking that it is hard to learn the science because of all the big words, look at it this way: While learning the science, you will learn some new languages!

✔ **QUICK CHECK**

From which languages do most biological terms originate?

Answer: Greek and Latin.

The More the Merrier . . . **Types of Terms**

When studying biology, you will encounter many different types of terms, many of which you will already know. Ideally, all terms used in biology would have the same origin and follow the same rules, but in reality, of course, that is simply not true.

DESCRIPTIVE TERMS

Most terms you will encounter are descriptive, meaning the name is closely related to its meaning. For most students, these words often look the most intimidating—they can be quite long. Yet, once you know some simple rules, **descriptive terms** become quite easy to understand and deciphering their meanings actually becomes fun. Even large words like *paleoanthropology* become manageable with some information and practice (look ahead to **Table 3.1**), and just think how empowered you will feel when you can actually converse with a scientist in his or her own language! We will spend most of our time in this chapter working on descriptive terms.

EPONYMS

Another type of term is the **eponym**, which literally means "putting a name upon." Eponyms are terms that include someone's name and have been used traditionally to honor the person who first described a certain structure or condition. This practice was particularly common in anatomy, the study of structures of organisms, which is one of the earlier branches of biology. This practice led to terms such as the

- Islets of Langerhans in your pancreas, first described by German pathological anatomist Paul Langerhans;

- Malpighian tubules, a system of excretory organs found in insects. These were named after Marcello Malpighi, a 17th-century anatomist; and
- Golgi apparatus, a structure with the specialized function of packaging and transporting substances within the cell. These were named for Camillo Golgi, an Italian cell biologist.

Obviously, these terms are not so easily understood. Few people now associate an eponym with the discoverer, making the terms harder to master. Recent practice in anatomy, and in all of biology, has moved away from eponyms to a preference for descriptive terms, so those will be our focus. You will learn relevant eponyms as you move through your course, but fewer and fewer are in use today.

Of course, there are exceptions. Down syndrome was previously referred to as *mongolism* because the characteristic eye appearance of someone with Down syndrome was a bit like that of people of Asian descent. This term, however, is deemed offensive, so the eponym is now used more often. Another descriptive term for this genetic disorder is trisomy 21, which tells us that three copies of chromosome 21 cause this condition. Although few people know what trisomy 21 means, most people recognize and understand the eponym.

Some of the terms for your course are part of a broader area known as *medical terminology*. Eponyms are still used relatively commonly in medical terminology and in health-related fields, especially for naming diseases or abnormal conditions. You will occasionally encounter them, and you really just need to memorize them as they come up.

TIME TO TRY

Parkinson's disease is also called *paralysis agitans*.

1. Which is the descriptive term? ______________________
2. Which is the eponym? ______________________
3. With which term are you more familiar? ______________________

►

4. The first vertebra of your spine, located just under your skull, is also called the *atlas*, which is an eponym. Either recall who Atlas was in Greek mythology, or look up his name. How is this eponym also a descriptive term? ______________________

____

Although paralysis agitans is the descriptive term, you are likely more familiar with the eponym—Parkinson's disease. You also likely realize that the atlas (bone) holds up your head, similar to the way Atlas from mythology held up the world.

ABBREVIATIONS AND ACRONYMS

Like many disciplines, the world of biology is full of shortcuts. After all, some of those descriptive terms grow quite large! You will often encounter modifications of full terms in the form of abbreviations or acronyms. You are already familiar with abbreviations. An **abbreviation** is a shortened form of a word or phrase. For example, the part of your digestive system that includes your stomach and intestines is referred to as the *gastrointestinal tract*, often abbreviated as the GI tract. Valves inside your heart regulate the blood flow between your heart's upper chambers, called atria, and its lower chambers, called ventricles. These valves are known as *atrioventricular valves*, which is often shortened to AV valves.

An **acronym** is technically a word formed from the first or key letters of each word in the full name. It is pronounced as if it is a word. For example, AIDS is the acronym for Acquired Immune Deficiency Syndrome and is pronounced like the word "aids." Sometimes regular abbreviations are formed like an acronym by using only certain letters from multiple words. For example, HIV stands for human immunodeficiency virus, the cause of AIDS. But we pronounce the acronym, AIDS, as a word, while spelling out the abbreviation—H-I-V. Many sources lump abbreviations and acronyms together, and surely acronyms are a special type of abbreviation, but now you know the difference.

Abbreviations and acronyms can be problematic, however. SAD is an acronym for a condition called *seasonal affective disorder*. This disorder can be characterized as serious winter blues, apparently brought on by decreased daylight in winter months. The depression that accompanies it can be severe, making the acronym particularly suitable. However, a search of the medical abbreviations section of Medilexicon.com identified 21 terms that match this abbreviation, and even more can be found at other sites. They include the following:

- Separation Anxiety Disorder,
- Social Anxiety Disorder,
- Small Airway Disease,
- Systemic Autoimmune Diseases, and
- Sporadic Alzheimer's Disease.

As you see, it is important to know the full name as well as the abbreviation, and to consider the context of how an abbreviation is used when determining its meaning. For example, an abbreviation used in medicine is AAA, referred to as a "Triple A," which stands for abdominal aortic aneurysm (**Figure 3.2**), a condition often seen in cadavers in

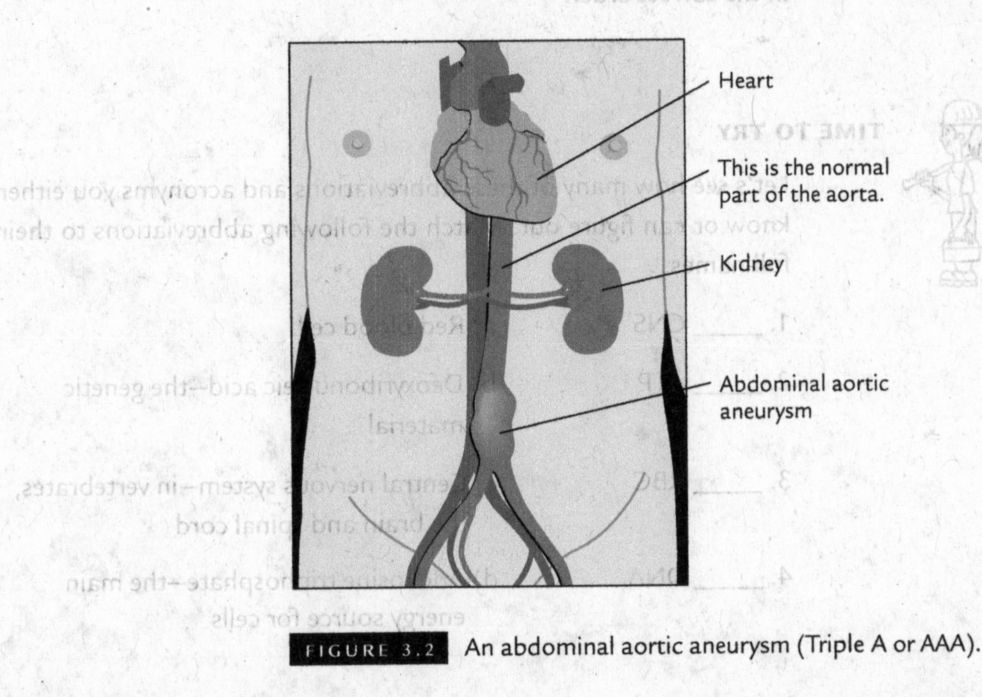

FIGURE 3.2 An abdominal aortic aneurysm (Triple A or AAA).

an anatomy lab. This is a weakening and ballooning of the aorta, the body's largest artery, where it ends in the abdomen. If it ruptures, it can kill you quickly. But if you are en route to a meeting of AAA (the American Association of Anatomists) and your tire ruptures, Triple A or AAA—the American Automobile Association—will come and fix it!

Here are two last cautions about abbreviations: There may be more than one for the same thing, and some that sound quite similar may be different. For example, a *CAT* scan is an image created through a technology called computer-aided tomography, but these are now more commonly called *CT* scans. An *ECG* and an *EKG* are the same thing. Both abbreviations stand for electrocardiogram, a tracing of your heart's electrical activity—it doesn't matter if you use a "C" or a "K." But, if you change the middle letter to an "E," you are referring to an electroencephalogram, which is a tracing of your brain's electrical activity. Use abbreviations with caution, and don't worry much about them now—like eponyms, you will learn them as you go and it is easier to do so.

When using abbreviations and acronyms, always be sure that you know the full names and their meanings, and that you list the letters in the correct order. ■

TIME TO TRY

Let's see how many of these abbreviations and acronyms you either know or can figure out. Match the following abbreviations to their full names.

1. _____ CNS — a) Red blood cell
2. _____ ATP — b) Deoxyribonucleic acid–the genetic material
3. _____ RBC — c) Central nervous system–in vertebrates, the brain and spinal cord
4. _____ DNA — d) Adenosine triphosphate–the main energy source for cells

These abbreviations probably gave you little trouble—you just had to match the letters in the name with the abbreviation. CNS is the central nervous system, ATP stands for the energy molecule adenosine triphosphate, RBC is a red blood cell, and DNA is the genetic material deoxyribosenucleic acid.

✔ **QUICK CHECK**

1. What makes an acronym different from a regular abbreviation? __________

2. SIDS stands for Sudden Infant Death Syndrome. Is this an abbreviation or an acronym? __________

Answers: 1. Unlike a regular abbreviation, an acronym is a word formed by key letters from the multiple words that it represents. 2. SIDS is an acronym, pronounced as it is spelled.

Putting Down Roots and **Building Descriptive Terms**

Most biological terms are descriptive and are built from two or more of three basic parts:

- a **prefix**, at the beginning of the word,
- a **root**, which is the main focus of the word, and
- a **suffix**, at the end of the word.

New words may be made simply by changing these parts (**Figure 3.3**). Some time ago, when learning English, you learned about prefixes and suffixes and how to use them to change one word to another, and you now use them all the time. You will be amazed how many word roots, prefixes, and suffixes you already know.

A prefix is a short addition placed at the front of a word root. Consider the root *cycle*, which means circle or wheel. We can change the specific meaning of this root by adding prefixes as follows:

- uni*cycle*—has one wheel
- bi*cycle*—has two wheels

- tri*cycle*—has three wheels
- motor*cycle*—has a motor

You likely knew the prefixes we just used in our example. When you are reading biological terms, don't let the first glance worry you. Simply take a breath and look carefully at the word. Find the root, then examine any prefixes or suffixes that are used with it. Sometimes you may not know a particular word part at first, but you may figure it out by thinking of words you *do* know that include that part. Let's try some.

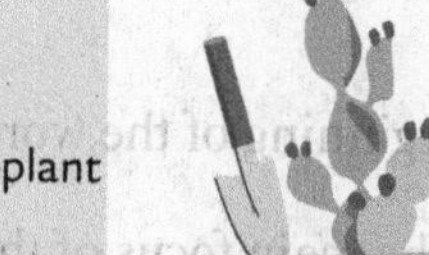

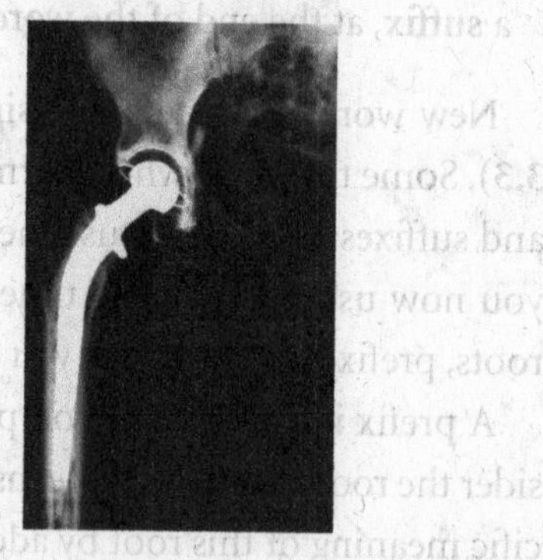

FIGURE 3.3 Adding prefixes and suffixes to the root "plant" will change the meaning.

TIME TO TRY

Here are some word parts and examples of words that use them. For each of the following, what do you think the word part means? Look at the examples and see if there is something they all share in common. This technique can help you determine the meaning of words with which you are unfamiliar—first look for the word parts and think of other terms that you do know that contain them.

Word part	Examples	Meaning
Root: cardi-	Cardiology, pericardium, cardiac arrest	
Prefix: multi-	Multicellular, multitude, multilingual	
Suffix: -ologist	Cardiologist, gynecologist, biologist	

From this exercise, you probably can see that *cardi-* refers to the heart, *multi-* means many or multiple, and *-ologist* means one who is knowledgeable in or studies whatever the root is. *Cardiology* is the study of the heart, and one who studies the heart is called a *cardiologist.* Note how similar these terms are, yet their meanings are different. The *pericardium* is the sac that surrounds the heart (*peri-* means *around*). And *cardiac arrest* refers to when the heart abruptly stops functioning, such as from a heart attack.

Let's consider the name of your current course: biology. The root *bio-* means "life," and the suffix *-ology* means "the study of" or "a branch of knowledge." Biology is therefore the study of the branch of knowledge pertaining to life. Now let's look at the name of another branch of biology, anatomy. The prefix *ana* means up or apart, and the root *tome* means a cutting. *Anatomy* is a combination of the two. It is the science that focuses on the parts and organization of the body. But, when early folks tried to discover how we are built, all they saw was the outside. To discover how the body is organized, they did dissections—they *cut up* the bodies, *cutting away* outer structures to see deeper structures (**Figure 3.4**). See how descriptive the term is?

FIGURE 3.4 Early anatomists explored the body's structures by cutting away the outer structures to see inside.

Not only do descriptive terms convey a clear meaning, but they are easy to translate once you know the prefixes, suffixes, and roots. You already know several of these, and you will learn plenty more as you go. To help jump-start your language acquisition, **Table 3.1** lists many word parts commonly used in biology, along with their meanings and an example of each. Don't worry about learning them all now—they are there to help you start building your language skills. If you get stuck on a word that appears in your textbook or that your instructor uses in a lecture, use this table as a reference. You should read through the terms and see which ones you already recognize.

Mastering word roots, prefixes, and suffixes is the most effective way to expand your biology vocabulary because they can be recombined into countless new words. ■

TABLE 3.1 Some common word parts used in biology.

Word part	Meaning	Example
a-	not, without	asexual reproduction: a type of reproduction involving only one parent that produces genetically identical offspring
abyss-	deep, bottomless	abyssal zone: the cold, dark benthic communities near the bottom of the ocean
acro	tip	acrosomal reaction: the discharge of a special organelle at the tip of a sperm called an acrosome that facilitates fertilization when the sperm approaches an egg
adeno-	gland	adenohypophysis: a gland, also called the anterior pituitary, positioned at the base of the hypothalamus
aero-	air	aerobic: referring to chemical reactions using oxygen
agglutinat-	glued together	agglutination: an immune response in which bacteria or viruses are clumped together
agon	a contest	agonistic behavior: a type of behavior involving a contest of some kind that determines which competitor gains access to some resource, such as food or mates
aktin	a ray	Actinopterygii: the class of rayfinned fishes
-alexi	to ward off	phytoalexin: an antibiotic, produced by plants, that destroys microorganisms or inhibits their growth
allo-	different	allopatric speciation: the formation of a new species due to a subpopulation becoming isolated from the parent population by a geographical barrier
alveol-	a cavity	alveoli: one of the dead-end, multilobed air sacs that constitute the gas exchange surface of the lungs

▶

TABLE 3.1 Some common word parts used in biology, continued.

Word part	Meaning	Example
ama	together	amacrine cell: nerve cells of the retina in the eye that help integrate information before it is sent to the brain
amphi-	dual	amphipathic molecule: a molecule that has both a hydrophobic and a hydrophilic region
an-	without, not	anion: a negatively charged ion
ana-	up	anabolic pathway: a metabolic pathway that consumes energy to build complex molecules from simpler ones
analog-	proportion	analogous: referring to features of organisms that are superficially similar but have evolved in different ways due to similar habitats
andro	male	polyandry: a polygamous mating system involving one female and many males
aneu-	without	aneuploidy: a chromosomal abnormality in which certain chromosomes are present in extra copies or are deficient in number
-angio	vessel	gametangia: the reproductive organ of bryophytes; a multichambered jacket of sterile cells in which gametes are formed
annus, -ennis	year	perennial: a plant that lives for many years
anth-	a flower	anther: in the flower of flowering plants, the terminal pollen sac of a stamen, inside which pollen grains with male gametes form
anthrop	man	paleoanthropology: the study of human origins and evolution
anti	against	anticodon: a specialized base triplet on one end of a tRNA molecule that recognizes a particular complementary codon on an mRNA molecule antibody: a protein that attaches to a specific kind of antigen and helps counter its effects; part of an immune response
-aphy	suck	anaphylactic shock: an acute, life-threatening allergic response
apic	tip	apical meristem: actively dividing plant tissue in the tips of roots and in the buds of shoots that produces cells that allow the plant to grow in length
apo-	off, away	apoplast: in plants, an extracellular pathway provided by the continuous matrix of cell walls
-apsis	juncture	synapsis: the pairing of replicated homologous chromosomes during prophase I of meiosis
aqua	water	aqueous humor: the clear, watery solution that fills the anterior cavity of the eye
arachn	spider	Arachnida: the arthropod group that includes scorpions, spiders, ticks, and mites

TABLE 3.1 Some common word parts used in biology, continued.

Word part	Meaning	Example
arch	ancient, beginning	archosaurs: the reptilian group that includes crocodiles, alligators, dinosaurs, and birds
arthro	jointed	Arthropoda: segmented invertebrates with exoskeletons and jointed appendages
astro-	a star	astrocytes: cells that provide structural and metabolic support for nerve cells
atrio	a vestibule	atrioventricular node: a region of specialized heart muscle tissue; it generates electrical impulses that cause the ventricles to contract
auto-	self	autonomic nervous system: the branch of the peripheral nervous system of vertebrates that regulates the internal environment
aux-	grow, enlarge	auxins: a class of plant hormones having a variety of effects, such as stimulation of cell elongation and secondary growth, and the development of fruit
bacul	a rod	baculum: a bone that is contained in, and helps stiffen, the penis of rodents, raccoons, walruses, and several other species of mammals
bentho-	the depths of the sea	benthic zone: the bottom surfaces of aquatic environments
bi	two	bipolar cell: nerve cells that synapse (communicate) with the axons of rods and cones in the retina of the eye
bin-	two at a time	recombinant: an offspring whose phenotype differs from that of the parents
bio-, biot-	life	biology: the scientific study of life biosphere: all the environments on Earth that are inhabited by life antibiotic: a chemical that kills bacteria or inhibits their growth
-bios	life	symbiosis: an ecological relationship between organisms of two different species that live together in direct contact
-blast	bud, sprout	osteoblasts: bone-forming cells
blasto	produce	blastomere: small cells of an early animal embryo
brachio	an arm	brachiopod: also called lamp shells, these marine animals superficially resemble clams and other bivalve molluscs, but the two halves of the brachiopod shell are dorsal and ventral to the animal rather than lateral, as in clams
bryo	moss	bryozoan: colonial animals (phylum Ectoprocta) that superficially resemble mosses
cala	insert	intercalated disks: specialized junctions between cardiac muscle cells that provide direct electrical coupling among cells
capsa	a box	capsid: the protein shell that encloses the viral genome

▶

TABLE 3.1 Some common word parts used in biology, continued.

Word part	Meaning	Example
carb-	coal	carboxyl group: a functional group present in organic acids, consisting of a carbon atom double-bonded to an oxygen atom and a hydroxyl group
cardi-	heart	cardiovascular system: the closed circulatory system characteristic of vertebrates
-carp	fruit	pericarp: the thickened wall around a fruit
carp-	fruit	carpel: the female reproductive organ of a flower, consisting of the stigma, style, and ovary
cata-	down	catabolic pathway: a metabolic pathway that releases energy by breaking down complex molecules into simpler ones
-centesis	a puncture	amniocentesis: a technique for determining genetic abnormalities in a fetus by the presence of certain chemicals or defective fetal cells in the amniotic fluid, obtained by aspiration from a needle inserted into the uterus
centro-	center	centromere: the narrow "waist" of a condensed replicated chromosome, where two sister chromatids are attached
cephal, cephalo	head	cephalization: an evolutionary trend toward the concentration of sensory equipment on the anterior end of the body cephalochordates: a chordate without a backbone, represented by lancelets, tiny marine animals
cheli	a claw	chelicerae: clawlike feeding appendages characteristic of the cheliceriform group of invertebrate animals
chemi-	chemical	chemiosmosis: the production of ATP using the energy of hydrogen ion gradients across membranes to phosphorylate ADP
chemo	chemical	chemoreceptor: a receptor that transmits information about the total solute concentration in a solution or about individual kinds of molecules
chiasm-	marked crosswise	chiasma: the X-shaped, microscopically visible region representing homologous chromosomes that have exchanged genetic material through crossing over during meiosis
chloro-	green	chloroplast: the site of photosynthesis in plants and eukaryotic algae
choano	a funnel	choanocyte: flagellated collar cells of a sponge
-chol	anger	catecholamines: a class of neurotransmitters and hormones, including epinephrine and norepinephrine, that are synthesized from the amino acid tyrosine
chondro-	cartilage	chondrocytes: cartilage cells
chord	string	notochord: a long flexible rod that runs along the dorsal axis of the body in all chordate animals

TABLE 3.1 Some common word parts used in biology, continued.

Word part	Meaning	Example
chroma-	colored	chromatin: DNA and the various associated proteins that form eukaryotic chromosomes
chrom	colored	cryptochrome: the name given to the unidentified blue-light photoreceptor
chylo	juice	chylomicron: small globules composed of fats that are mixed with cholesterol and coated with special proteins for transport through the blood
chymo	juice	chymotrypsin: an enzyme found in the small intestine of vertebrates
cili-	hair	cilium: a short hair-like cellular appendage with a microtubule core
circ-, circa-	circle	circadian rhythm: a physiological cycle of about 24 hours, present in all eukaryotic organisms, that persists even in the absence of external cues
clado-	branch	cladogram: a dichotomous phylogenetic tree that branches repeatedly
-clin	slope	thermocline: a narrow stratum of rapid temperature change in the ocean and in many temperate-zone lakes
cnido	a nettle	cnidocytes: unique cells that function in defense and prey capture in the invertebrate animals called cnidarians
co-	together	codominance: a situation in which two alleles in the heterozygous genotype of an organism are fully expressed in its phenotype
coch	a snail	cochlea: the complex, coiled organ of hearing that contains the organ of Corti
coel	hollow	spongocoel: the central cavity of a sponge
coeno	common	coenocytic: a multinucleated condition resulting from the repeated division of nuclei without cytoplasmic division
coit	a coming together	coitus: sexual intercourse
coleo-	a sheath	coleorhiza: the covering of the young root of the embryo of a grass seed
coll-	glue	collenchyma cell: a flexible plant cell type that occurs in strands or cylinders that support young parts of the plant without restraining growth
com-	together	recombinant: an offspring whose phenotype differs from that of the parents
con-	with, together	condensation reaction: a chemical reaction in which two molecules become covalently bonded to each other through the loss of a small molecule, usually water
conjug	together	conjugation: in bacteria, the transfer of DNA between two cells that are temporarily joined
contra	against	contraception: the prevention of pregnancy

▶

TABLE 3.1 Some common word parts used in biology, continued.

Word part	Meaning	Example
cope	an oar	copepods: a group of small crustaceans that are important members of marine and freshwater plankton communities
cortex	shell	cortical reaction: a series of changes in the cortex of the egg cytoplasm during fertilization
-cortico	the shell	adrenocorticotropic hormone: a peptide hormone released from the anterior pituitary, it stimulates the production and secretion of steroid hormones by the adrenal cortex
counter-	opposite	countercurrent exchange: the opposite flow of adjacent fluids that maximizes transfer rates
crani	the skull	craniata: the chordate clade that possess a cranium
crypto-	hidden	cryptic coloration: a type of camouflage that makes potential prey difficult to spot against its background
-cutane	skin	pulmocutaneous: the route of circulation that directs blood to the skin and lungs
cuti	the skin	cuticle: the exoskeleton of an arthropod
-cycle	circle	pericycle: a layer of cells just inside the endodermis of a plant root that may become meristematic and begin dividing again, producing root branches
cyclo-	circle	cyclin: a regulatory protein whose concentration fluctuates cyclically
cyst	sac, bladder	blastocyst: a hollow ball of cells produced one week after fertilization in humans
-cyte	cell	chondrocytes: cartilage cells
cyto-	cell	cytosol: the semifluid medium in a cell in which organelles are located
de	from, down, out	denitrification: the process of converting nitrate back to nitrogen
deca	ten	decapod: a relatively large group of crustaceans characterized by having ten appendages; includes lobsters, crayfish, crabs, and shrimp
del	visible	Urodela: the order of salamanders that includes amphibians with tails
demo-	people	demography: the study of statistics relating to births and deaths in populations
dendro-	tree	dendrite: one of usually numerous, short, highly branched processes of a nerve cell that receive signals from other neurons
-derm	skin	endodermis: the innermost layer of the cortex in plant roots
-desma	a band or bond	plasmodesmata: an open channel in a plant cell wall
detrit-	wear off	detritivore: a consumer that derives its energy from nonliving organic material

TABLE 3.1 Some common word parts used in biology, continued.

Word part	Meaning	Example
deutero	second	deuterostome: one of two lines of coelomates characterized by radial, indeterminate cleavage, enterocoelous formation of the coelom, and development of the anus from the blastopore
di-	two	diploid: cells that contain two homologous sets of chromosomes
-dilat	expanded	vasodilation: an increase in the diameter of blood vessels, allowing more blood to flow through them
dino	terrible	dinosaurs: an extremely diverse group of ancient reptiles varying in body shape, size, and habitat
diplo	double	Diplopoda: the millipede group of animals, which are characterized by having two sets of legs on each body segment diploid: cells that contain two homologous sets of chromosomes
dis-	separate	nondisjunction: an accident of meiosis or mitosis, in which both members of a pair of homologous chromosomes or both sister chromatids fail to separate
diure	urinate	antidiuretic hormone: a hormone that helps regulate water balance
dorm-	sleep	dormancy: a condition typified by extremely low metabolic rate and a suspension of growth and development
ecdys-	an escape	ecdysone: a steroid hormone that triggers molting in arthropods
echino	spiny	echinoderm: sessile or slow-moving animals with a thin skin that covers an endoskeleton; the group includes sea stars, sea urchins, and sea cucumbers
-ecious	house	monoecious: referring to a plant species that has both staminate and carpellate flowers on the same individual
eco-	house	ecosystem: a biological community and the physical environment associated with it
ecto-	outside	ectoparasites: parasites that feed on the external surface of a host
ectomy	cut out	vasectomy: a surgical procedure involving the cutting of each vas deferens to prevent sperm from entering the urethra
electro	electricity	electroporation: a technique used to introduce recombinant DNA into cells by applying a brief electrical pulse to a solution containing those cells
-ell	small	organelle: a small formed body with a specialized function found in the cytoplasm of eukaryotic cells
encephal	brain	electroencephalograph: instrument for measuring and recording the electrical activity of the brain
enanti-	opposite	enantiomer: molecules that are mirror images of each other

▶

TABLE 3.1 Some common word parts used in biology, continued.

Word part	Meaning	Example
-enchyma	an infusion	collenchyma cell: a flexible plant cell type that occurs in strands or cylinders that support young parts of the plant without restraining growth
end-, endo	within, inner	endemic: a type of species that is found only in one region and nowhere else in the world endoskeleton: a hard skeleton buried within the soft tissues of an animal
entero	intestine, gut	enterogastrones: a category of hormones secreted by the wall of the small intestine in vertebrates
epi-	above, over	epidermis: the dermal tissue system in plants; the outer covering of animals
erythro-	red	erythropoietin: a hormone produced in the kidney that stimulates the production of erythrocytes erythrocytes: red blood cell
estuar-	the sea	estuary: the area where a freshwater stream or river merges with the ocean
etho	custom, habit	ethology: the study of animal behavior in natural conditions
eu-	true, good	eukaryotic cell: a cell that has a true nucleus
eury	broad, wide	eurypterid: an extinct group of mainly marine and freshwater cheliceriforms; these predators, also called water scorpions, ranged up to 3 meters long
ex-, exo-	out, outside	exergonic reaction: a reaction that proceeds with a net release of free energy exocytosis: the movement of materials out of a cell
extero	outside	exteroreceptor: sensory receptors that detect stimuli outside the body, such as heat, light, pressure, and chemicals
extra	outside	extracellular digestion: the breakdown of food outside cells
fertil	fruitful	fertilization: the union of haploid gametes to produce a diploid zygote
fibrino-	a fiber	fibrinogen: the inactive form of the plasma protein that is converted to the active form fibrin, which aggregates into threads that form the framework of a blood clot
fibro-	fiber	fibroblast: a type of cell in loose connective tissue that secretes the protein ingredients of the extracellular fibers
flagell-	whip	flagellum: a long whip-like cellular appendage that moves cells
-form	shape	transformation: the process that converts a normal cell into a cancer cell
fovea	a pit	fovea: the center of the visual field of the eye
fusi-	a spindle	fusiform initials: the cambium cells within the vascular bundles of a flowering plant; the name refers to the tapered ends of these elongated cells

TABLE 3.1 Some common word parts used in biology, continued.

Word part	Meaning	Example
gam-	marriage	monogamous: a type of relationship in which one male mates with just one female
gamet-	wife or husband	gamete: a haploid egg or sperm cell
gamy	reproduction, marriage	syngamy: the process of cellular union during fertilization
gastro	stomach, belly	gastrulation: in animal development, the formation of a gastrula, which is a three-layered embryo with a primitive gut
gen-, gene-	produce	genome: a cell's endowment of DNA polygenic: an additive effect of two or more gene loci on a single phenotypic character
genesis	origin, birth	in vitro mutagenesis: a technique to discover the function of a gene by introducing specific changes into the sequence of a cloned gene, reinserting the mutated gene into a cell, and studying the phenotype of the mutant
-genic	producing	electrogenic pump: an ion transport protein that generates voltage across a membrane
geno-	offspring	genotype: the genetic makeup of an organism
-geny	origin	phylogeny: the evolutionary history of a related group of organisms called a taxon
geo-	the Earth	biogeography: the study of the past and present distribution of species
gest	carried	ingestion: a heterotrophic mode of nutrition in which other organisms or detritus are eaten
gibb-	humped	gibberellins: a class of related plant hormones that stimulate growth in the stem and leaves, trigger the germination of seeds and breaking of bud dormancy, and stimulate fruit development with auxin
glia	glued together	glia: supporting cells that are essential for the structural integrity of the nervous system and for the normal functioning of nerve cells
-glob	globe, sphere	immunoglobulin: one of the class of proteins that act as antibodies in an immune response
glomer	a ball	glomerulus: a ball of capillaries surrounded by Bowman's capsule in the nephron and serving as the site of filtration in the vertebrate kidney
glotti	the tongue	epiglottis: a cartilaginous flap that blocks the opening of the windpipe into the pharynx, the glottis, during swallowing
gluco-	sweet	glucagon: a peptide hormone secreted by pancreatic endocrine cells that raises blood glucose levels; an antagonistic hormone to insulin
glyco-	sweet	glycogen: a polysaccharide sugar used to store energy in animals
gnatho	jaws	gnathostomes: the vertebrate clade that possesses jaws

▶

TABLE 3.1 Some common word parts used in biology, continued.

Word part	Meaning	Example
-graphy	writing	demography: the study of statistics relating to births and deaths in populations
gusta	taste	gustatory receptors: taste receptors
gutt-	a drop	guttation: the sweating of water droplets from leaves that is caused by root pressure in certain plants
gymno-	naked	gymnosperm: a vascular plant that bears naked seeds not enclosed in any specialized chambers
gyno	a woman	polygyny: a polygamous mating system involving one male and many females
halin	salt	euryhaline: organisms that can tolerate substantial changes in external osmolarity
halo-	salt	halophile: microorganisms that live in highly saline environments, such as the Great Salt Lake or the Dead Sea
haplo-	single	haploid: cells that contain only one chromosome of each homologous pair
helic-	a spiral	helicase: an enzyme that untwists the double helix of DNA at the replication forks
hemo-	blood	hemophilia: a human genetic disease caused by a sex-linked recessive allele, characterized by excessive bleeding following injury
herb	grass	herbivore: a heterotrophic animal that eats plants
hermaphrod	both male and female	hermaphrodite: an individual that functions as both male and female in sexual reproduction by producing both sperm and eggs
hetero-	other, different	heterozygous: an organism or cell that has two different alleles at a given gene location on homologous chromosomes
holo	whole	holoblastic cleavage: a type of cleavage in which the egg is completely divided
homeo-	same	homeostasis: maintenance of a stable internal environment of the body
homin	man	hominin: a term that refers to mammals that are more closely related to humans than to any other living species
homo-	alike	homozygous: an organism or cell that has two identical alleles at a given gene locus on homologous chromosomes
horm-	to set in motion, excite	hormone: in multicellular organisms, one of many types of circulating chemical signals
hydro-	water	hydrocarbon: an organic molecule consisting only of carbon and hydrogen
hyl	matter	mesohyl: a gelatinous region between the two layers of cells of a sponge

TABLE 3.1 Some common word parts used in biology, continued.

Word part	Meaning	Example
hyper-	over, above, excessive	hypersensitive response: in flowering plants, a vigorous, localized defense response to a pathogen
hypo-	under, below, lower	hypothalamus: the underside of the vertebrate forebrain that functions in maintaining homeostasis, especially in coordinating the endocrine and nervous systems; it is located below the thalamus
ichthy	fish	Osteichthyans: the vertebrate clade that includes the rayfinned fishes and lobefins
immuno-	safe, free	immunization: the process of generating a state of immunity to disease by artificial means
in	without	invertebrates: animals without a backbone
inter-	between	interphase: time when a cell metabolizes and performs its various functions before and after it divides
intero	inside	interoreceptor: sensory receptors that detect stimuli within the body, such as blood pressure and body position
intra	inside, within	intracellular digestion: the joining of food vacuoles and lysosomes to allow chemical digestion to occur within the cytoplasm of a cell
intro	within	intron: a noncoding, intervening sequence within a eukaryotic gene
involu	wrapped up	involution: the turning or rolling of cells over the edge of a lip into the interior that occurs during development of some vertebrates
-iodo	violet	triiodothyrodine: one of two very similar hormones produced by the thyroid gland
iso-	equal	isotope: one of two or more atoms of the same element having the same number of protons and electrons but a different number of neutrons
itero-	to repeat	iteroparity: a life history in which adults produce large numbers of offspring over many years; also known as repeated reproduction
-itis	inflammation	hepatitis: inflammation or infection of the liver
juxta	near to	juxtaglomerular apparatus: a specialized tissue located near the arteriole that supplies blood to the glomerulus of the kidney
karyo-	nucleus	karyotype: a display of the chromosomes of a cell arranged in pairs
kilo-	a thousand	kilocalorie: a thousand calories
kine, kinet-	movement	kinesis: a change in activity rate in response to a stimulus kinetic energy: the energy of motion
koilos	a hollow	acoelomate: the condition of lacking a coelom or inner body cavity
-kytos	vessel	phagocytosis: a form of cell eating in which a cell engulfs a smaller organism or food particle
labi	lip	labia majora: a pair of thick, fatty ridges that enclose and protect the human female genitalia

▶

TABLE 3.1 Some common word parts used in biology, continued.

Word part	Meaning	Example
lact	milk	lactation: the production of milk
lamin-	sheet, layer	nuclear lamina: a netlike array of protein filaments that maintains the shape of the nucleus in a cell
lepido	a scale	lepidosaurs: the reptilian group which includes lizards, snakes, and tuatara
leuko-	white	leukocyte: a white blood cell
liga-	bound or tied	ligand: a small molecule that specifically binds to a larger one
limn-	a lake	limnetic zone: the well-lit, open surface waters of a lake farther from shore
-lite	first	stromatolite: rocklike structure composed of layers of prokaryotes and sediment
-lith	stone	statolith: specialized organelles called plastids that help a plant tell up from down
littor-	seashore	littoral zone: the shallow, well-lit waters of a lake close to shore
lobi	a lobe	trilobite: an extinct group of arthropods with pronounced segmentation
lopho	crest, tuft	lophophore: a horseshoe-shaped or circular fold of the body wall bearing ciliated tentacles that surround the mouth in some invertebrate groups
-luna	moon	semilunar valve: a valve located at the two exits of the heart that resembles a crescent or half-moon
lut-	yellow	luteinizing hormone: a gonadotropin secreted by the anterior pituitary
-lyse	break	hydrolyse: to break chemical bonds by adding water
-lysis	split	glycolysis: the splitting of glucose into pyruvate
lyso-	loosen	lysosome: a membrane-bounded sac of hydrolytic enzymes that a cell uses to digest macromolecules
lyto	loosen	lytic cycle: a type of viral replication cycle resulting in the release of new viruses by death or lysis of the host cell
macro-	large	macromolecule: a large molecule
magnet-	magnetic	electromagnetic spectrum: the entire spectrum of radiation
major	larger	labia majora: a pair of thick, fatty ridges that enclose and protect the human female genitalia
mal-	bad or evil	malignant tumor: a cancerous tumor that is invasive enough to impair the functions of one or more organs
mamm-	breast or teat	mammal: any vertebrate of the class Mammalia that is characterized by having hair and the ability to produce milk from mammary glands to nurse young

TABLE 3.1 Some common word parts used in biology, continued.

Word part	Meaning	Example
marsupi	a bag, pouch	marsupial: a mammal, such as a koala, kangaroo, or opossum, whose young complete their embryonic development inside a maternal pouch called the marsupium
mechano	an instrument	mechanoreceptor: a sensory receptor that detects physical deformations in the body's environment associated with pressure, touch, stretch, motion, and sound
mega-	large, great	megaspore: a large, haploid spore that can give rise to a haploid female gametophyte in flowering plants
meio-	less	meiosis: a variation of cell division that yields haploid daughter cells, which have half as many chromosomes as their diploid parent cell
melan-	black	melanin: any of a group of polymers that cause pigmentation in vertebrates
menstru	month	menstruation: the shedding of portions of the endometrium during a menstrual cycle in most female primates
mere	part	blastomere: small cells of an early embryo
meristo-	divided	apical meristems: actively dividing plant tissue in the tips of roots and in the buds of shoots that produces cells that allow the plant to grow in length
mero	part	meroblastic cleavage: a type of embryonic cleavage in which there is incomplete division of a yolk-rich egg, characteristic of avian development
meso	middle	mesoderm: the middle primary germ layer of an early animal embryo
meta-	boundary, turning point; change; with; between	metaphase: the mitotic stage in which the chromosomes are aligned in the middle of the cell, at the metaphase plate
-metron	measure	allometric growth: the variation in the relative rates of growth of various parts of the body, which helps shape the organism
micro-	small	microspore: a small, haploid spore that can give rise to a haploid male gametophyte in flowering plants
minor	smaller	labia minora: a pair of slender skin folds that enclose and protect the vestibule in human female genitalia
missi	send	transmission: the conduction of impulses to the central nervous system
mito-	a thread	mitosis: the process of nuclear division in eukaryotic cells, in which the threadlike replicated chromosomes are allocated equally to the two daughter cells produced
mono-	one	monosomic: a chromosomal condition in which a particular cell has only one copy of a chromosome, instead of the normal two; the cell is said to be monosomic for that chromosome

▶

TABLE 3.1 Some common word parts used in biology, continued.

Word part	Meaning	Example
morph, morpho-	form	metamorphosis: the resurgence of development in an animal larva that transforms it into a sexually mature adult morphogenesis: the development of body shape and organization
morul	a little mulberry	morula: a solid ball of blastomeres formed by early cleavage during animal development
muta-	change	mutation: a change in the DNA of a gene
mutu-	reciprocal	mutualism: a symbiotic relationship in which both the host and the symbiont benefit
myco	fungus	mycorrhizae: mutualistic associations of plant roots and fungi
myo	muscle	myotonia: increased muscle tension
-nema	thread	protonema: a mass of green, branched, one-cell-thick filaments produced by germinating moss spores
nemato	a thread	nematocysts: the stinging capsules in cnidocytes, unique cells that function in defense and capture of prey in the invertebrate group called cnidarians
nephri	kidney	metanephridium: in annelid worms, a type of excretory tubule
neuro-	nerve	neurohypophysis: also called the posterior pituitary, this endocrine gland is an outgrowth from underneath the brain
neutr, neutro-	neutral	neutron: a subatomic particle with a neutral electrical charge
noci	harm	nociceptor: pain receptors in the epidermis of the skin
nom-	name	binomial: a two-part latinized name of a species
non-	not	nondisjunction: an accident of meiosis or mitosis, in which both members of a pair of homologous chromosomes or both sister chromatids fail to separate
noto	back	notochord: a long flexible rod that runs along the dorsal axis of the body in chordate animals
nucle-, nucleo	nucleus	nucleoid: the region where the genetic material is concentrated in prokaryotic cells nucleosome: the basic beadlike unit of DNA packaging in eukaryotes
-oid	like, form	nucleoid: the region where the genetic material is concentrated in prokaryotic cells
olfact	smell	olfactory receptor: smell receptors
oligo-	small, scant	oligotrophic lake: a nutrient-poor, clear, deep lake with minimal growth of phytoplankton
ology	the study of	paleoanthropology: the study of human origins and evolution
omma	the eye	ommatidia: the visual units of the compound eye of arthropods and some worms

TABLE 3.1 Some common word parts used in biology, continued.

Word part	Meaning	Example
omni	all	omnivore: a heterotrophic animal that consumes both meat and plant material
onco	tumor	proto-oncogene: a normal cellular gene with the potential to cause cancer if it is altered to become an oncogene
oo	egg	oogenesis: the process in the ovary that results in the production of female gametes
opercul	a covering, lid	operculum: a protective flap that covers the gills of fishes
oscul	a little mouth	osculum: a large opening in a sponge that connects its central cavity to the environment
osis	a condition of	mycosis: the general term for a fungal infection
osmo-	pushing	osmosis: the diffusion of water across a selectively permeable membrane
osteo-	bone	osteoblasts: bone-forming cells
ostraco	a shell	ostracoderm: an extinct group of armored fishlike vertebrates
ovi	an egg	oviparous: referring to a type of development in which young hatch from eggs laid outside the mother's body
oxy-	sharp, acid	oxytocin: a hormone that induces contractions of the uterine muscles and causes the mammary glands to eject milk during nursing
paedo-	child	paedomorphosis: the retention in the adult organism of the juvenile features of its evolutionary ancestors
paleo-	ancient	paleontology: the scientific study of fossils
para-	beside, near	parasympathetic division: one of three divisions of the autonomic nervous system
-parity	to beget	semelparity: a life history in which adults have but a single reproductive opportunity to produce large numbers of offspring, such as the life history of the Pacific salmon; also known as "big-bang reproduction"
parous	bearing	oviparous: referring to a type of development in which young hatch from eggs laid outside the mother's body
parsi-	few	principle of parsimony: the premise that a theory about nature should be the simplest explanation that is consistent with the facts
partheno	without fertilization, virgin	parthenogenesis: a type of reproduction in which females produce genetically identical offspring from unfertilized eggs
partur	giving birth	parturition: the expulsion of a baby from the mother; also called birth
-patri	father	sympatric speciation: the formation of a new species as a result of a radical change in the genome that produces a reproductively isolated subpopulation in the midst of its parent population

TABLE 3.1 Some common word parts used in biology, continued.

Word part	Meaning	Example
pedi-	a child	pedigree: a family tree describing the occurrence of heritable characters in parents and offspring across as many generations as possible
pelag-	the sea	oceanic pelagic biome: most of the ocean's waters far from shore, constantly mixed by ocean currents
pellucid	transparent	zona pellucida: the clear extracellular matrix surrounding a mammalian egg
perfora-	bore through	perforin: a protein that forms pores in a target cell's membrane
peri	around	peritubular capillaries: the network of tiny blood vessels that surrounds the proximal and distal tubules in the vertebrate kidney
perman-	remaining	permafrost: a permanently frozen layer below the arctic tundra
-phage	to eat	bacteriophages: viruses that infect bacteria
phago-	to eat	phagocytosis: cell eating; the process by which food or foreign particles are enveloped by cells called phagocytes
pheno-	appear	phenotype: the physical and physiological traits of an organism
-phil	loving	neutrophil: the most abundant type of leukocyte, neutrophils destroy foreign invaders and release various substances
-philia	loving	hemophilia: a human genetic disease caused by a sex-linked recessive allele, characterized by excessive bleeding following injury
-philos	loving	hydrophilic: having an affinity for water
phloe-	the bark of a tree	phloem: the portion of the vascular system in plants consisting of living cells arranged into elongated tubes that transport sugar and other organic nutrients throughout the plant
-phobos	fearing	hydrophobic: having an aversion to water
phora	to carry	lophophore: a horseshoe-shaped or circular fold of the body wall bearing ciliated tentacles that surround the mouth in some invertebrate groups
-phore	bearer	gametophore: the mature gamete-producing structure of a gametophyte body of a moss
photo	light	photoreceptor: receptors of light
phragmo-	a partition	phragmoplast: an alignment of cytoskeletal elements and Golgi-derived vesicles across the midline of a dividing plant cell
-phyll	leaf	chlorophyll: photosynthetic pigment in chloroplasts that colors leaves and other plant parts
phylo-	tribe	phylogeny: the evolutionary history of a group of related organisms called a taxon
-phyte	plant	bryophytes: the mosses, liverworts, and hornworts; a group of nonvascular plants that inhabit the land but lack many of the terrestrial adaptations of vascular plants

TABLE 3.1 Some common word parts used in biology, continued.

Word part	Meaning	Example
phyto-	plant	phytoremediation: an emerging, nondestructive technology that seeks to cheaply reclaim contaminated areas by taking advantage of the remarkable ability of some plant species to extract heavy metals and other pollutants from the soil and to concentrate them in easily harvested portions of the plant
pino-	drink	pinocytosis: cell drinking; the process by which a living cell engulfs a droplet of liquid
placo	a plate	placoderm: a member of an extinct group of jawed fishes that were enclosed in a tough, outer armor
plan	flat or wandering	planarians: flatworms
plasm-, plasmo	molded	plasmolysis: a phenomenon in walled cells such as those of plants in which the cytoplasm shrivels and the plasma membrane pulls away from the cell wall when the cell loses water to a hypertonic environment
-plast	formed, molded	protoplast: the contents of a plant cell exclusive of the cell wall
pleio-	more	pleiotropy: when a single gene impacts more than one characteristic
pluri-	more, several	pluripotent stem cell: a cell within bone marrow that can give rise to any kind of blood cell
-pod, -podium	foot	pseudopodium: a cellular extension of amoeboid cells used in moving and feeding
		tetrapod: a terrestrial vertebrate possessing two pairs of limbs, such as amphibians, reptiles, and mammals
podo-	foot	podocytes: specialized cells of Bowman's capsule in the vertebrate kidney that are permeable to water and small solutes but not to blood cells or large molecules such as plasma proteins
-poiet	produce	erythropoietin: a hormone produced in the kidney that stimulates the production of erythrocytes
poly	many	polygamous: a type of relationship in which an individual of one sex mates with several of the other sex
-pomo	fruit	apomixis: the asexual production of seeds
pore	a passage	blastopore: during animal development, the opening of the archenteron in the gastrula that develops into the mouth in protostomes and the anus in deuterostomes
-pori	small opening	aquaporin: a transport protein in the plasma membrane of a plant or animal cell that specifically facilitates the diffusion of water across the membrane
-port	gate, door	cotransport: the coupling of the "downhill" diffusion of one substance to the "uphill" transport of another across a membrane against its own concentration gradient

TABLE 3.1 Some common word parts used in biology, continued.

Word part	Meaning	Example
post-	after	postsynaptic cell: the target cell at a synapse
potent	powerful	totipotent: referring to a cell with the ability to form all parts of a mature organism
pre-	before	presynaptic cell: the transmitting cell at a synapse
pro-	before	prophase: the first mitotic stage, in which the chromatin is condensing
profund-	deep	profundal zone: the deepest region of a lake, where no light penetrates
proto	first	protobionts: aggregates of inorganically produced molecules
pseudo	false	pseudogenes: DNA segments very similar to real genes but which do not yield functional products
pter-	a wing, a feather, a fin	pterophytes: seedless plants with true roots with vascular tissue; the group includes ferns, whisk ferns, and horsetails
ptero	a wing	pterosaurs: extinct winged reptiles that lived during the time of dinosaurs
pterygi	a fin	Actinopterygii: the class of rayfinned fishes
pulmo-	a lung	pulmocutaneous: the route of circulation that directs blood to the skin and lungs
quadr-	four	quadruped: any animal having four feet
radia	a spoke, ray	Radiata: the radially symmetrical animal phyla, including cnidarians
ratit	flatbottomed	ratites: the group of flightless birds
re-	again	recombinant: an offspring whose phenotype differs from that of the parents
recta	straight	vasa recta: the capillary system that serves the loop of Henle in the kidney
regula	regular	osmoregulation: adaptations to control the water balance in organisms living in hyperosmotic, hypoosmotic, or terrestrial environments
reni	a kidney	renin-angiotensin-aldosterone system: part of a complex feedback circuit that normally partners with antidiuretic hormone in osmoregulation
retro	backward	retrovirus: an RNA virus that reproduces by transcribing its RNA into DNA and then inserting the DNA into a cellular chromosome
rhizo-	root	rhizoids: long tubular single cells or filaments of cells that anchor bryophytes to the ground
rhodo	red	rhodopsin: a photosensitive purple-red protein in the rods of the retina in the eye
sacc	a sack	saccule: a chamber in the inner ear that participates in the sense of balance
-sacchar	sugar	monosaccharide: simplest type of sugar

TABLE 3.1 Some common word parts used in biology, continued.

Word part	Meaning	Example
salta-	leap	saltatory conduction: rapid transmission of a nerve impulse along an axon resulting from the action potential jumping from one node of Ranvier to another, skipping the myelin-sheathed regions of membrane
sarco	flesh	sarcomere: the fundamental, repeating unit of striated muscle, delimited by the Z lines
saur	lizard	dinosaurs: an extremely diverse group of ancient reptiles varying in body shape, size, and habitat
schizo	split	schizocoelous: the type of development found in protostomes; initially, solid masses of mesoderm split to form coelomic cavities
sclero	hard	sclera: a tough, white outer layer of connective tissue that forms the globe of the vertebrate eye
script	write	transcription: the synthesis of RNA on a DNA template
scutell-	a little shield	scutellum: a specialized type of seed leaf found in the grass family
semel-	once	semelparity: a life history in which adults have but a single reproductive opportunity to produce large numbers of offspring, such as the life history of the Pacific salmon; also known as "big-bang reproduction"
semi	half	semicircular canals: a three-part chamber of the inner ear that functions in maintaining equilibrium
simi	an ape	prosimians: a suborder of primates that probably resemble early arboreal primates
socio	a companion	sociobiology: the study of social behavior based on evolutionary theory
soma	body	acrosome: an organelle at the tip of a sperm cell that helps the sperm penetrate the egg centrosome: a nonmembranous organelle that functions throughout the cell cycle to organize the cell's microtubules
-sperm	seed	gymnosperm: a vascular plant that bears naked seeds not enclosed in any specialized chambers
sporo-	seed	sporophyte: the multicellular diploid form in organisms such as plants that undergo alternation of generations that results from a union of gametes and that meiotically produces haploid spores that grow into the gametophyte generation
stalsis	constriction	peristalsis: rhythmic waves of contraction of smooth muscle that push food along the digestive tract
stam-	standing upright	stamen: the pollen-producing male reproductive organ of a flower, consisting of an anther and filament
-stasis	standing, staying	epistasis: a phenomenon in which one gene alters the expression of another gene that is independently inherited

▶

TABLE 3.1 Some common word parts used in biology, continued.

Word part	Meaning	Example
stato-	standing, placed	statolith: specialized organelles called plastids that help a plant tell up from down
steno	narrow	stenohaline: referring to organisms that cannot tolerate substantial changes in external osmolarity
stoma	mouth	protostomes: a member of one of two distinct evolutionary lines of coelomates characterized by spiral, determinate cleavage, schizocoelous formation of the coelom, and development of the mouth from the blastopore
stromato	something spread out	stromatolite: rocklike structure composed of layers of prokaryotes and sediment
sulf-	sulfur	sulfhydryl group: a functional group that consists of a sulfur atom bonded to an atom of hydrogen
supra-	above, over	suprachiasmatic nuclei: a pair of structures in the hypothalamus of mammals that functions as a biological clock
sym-	with, together	symplast: in plants, the continuum of cytoplasm connected by plasmodesmata between cells
-symmetr	symmetrical	asymmetric cell division: cell division in which one daughter cell receives more cytoplasm than the other
syn-	together	synapse: the locus where a neuron communicates with a postsynaptic cell in a neural pathway
		synthesis: combining separate elements into a single, unified whole
taxo-	arrange	taxonomy: the branch of biology concerned with naming and classifying the diverse forms of life
telos-	an end	telomere: the protective structure at each end of a eukaryotic chromosome
tens	stretched	renin-angiotensin-aldosterone system: part of a complex feedback circuit that normally partners with antidiuretic hormone in osmoregulation
tetan	rigid, tense	tetanus: the maximal, sustained contraction of a skeletal muscle
tetra-	four	tetrapod: a vertebrate with two pairs of limbs, including mammals, amphibians, and birds and other reptiles
thallos-	sprout	thallus: a seaweed body that is plantlike but lacks true roots, stems, and leaves
theca	a cup, case	spermatheca: a sac in the reproductive system of some female or hermaphrodite animals where sperm are stored
therm-	heat	thermodynamics: the study of the energy transformations that occur in a collection of matter
thermo	heat, temperature	thermoreceptor: an interoreceptor stimulated by either heat or cold

TABLE 3.1 Some common word parts used in biology, continued.

Word part	Meaning	Example
thigmo-	a touch	thigmomorphogenesis: a growth response in plants to chronic mechanical stimulation; an example is thickened stems in response to strong winds
thio-	sulfur	thiol: organic compounds containing sulfhydryl groups
thrombo-	a clot	thrombus: a clump of platelets and fibrin that blocks the flow of blood through a blood vessel; a blood clot
thylaco-	sac or pouch	thylakoid: a series of flattened saclike membranous structures within chloroplasts
tono-	stretched	tonoplast: the membrane that encloses a large central vacuole in a mature plant cell
-tonus	tension	hypertonic: referring to a solution with a higher concentration of solutes in comparison to another solution
-topo	place	epitope: a small region on the surface of an antigen to which an antigen receptor or antibody binds
toti	all	totipotent: referring to a cell with the ability to form all parts of a mature organism
trachei-	the windpipe	tracheids: elongated, perforated cells, which are dead at maturity, that make up part of the pipelike water-conducting xylem tissue in vascular plants
trans-	across	translocation: attachment of a chromosomal fragment to a nonhomologous chromosome
tri	three	trimester: a three-month period
trocho	wheel	trochophore: a ciliated larva common to the life cycle of many molluscs, it is also characteristic of marine annelids and some other groups
-trop	turn	phototropism: growth of a plant shoot toward or away from light
-troph	food, nourishment	photoautotroph: an organism that harnesses light energy to drive the synthesis of organic compounds from carbon dioxide
tropho	food, nourishment	trophoblast: the outer epithelium of the blastocyst, which forms the fetal part of the placenta in mammalian development
-tropic	to turn or change	adrenocorticotropic hormone: a hormone released from the anterior pituitary, which stimulates the production and secretion of hormones by the adrenal cortex
tropo	turn, change	tropomyosin: a regulatory protein that blocks muscle contraction
trypsi	wearing out	chymotrypsin: an enzyme found in the small intestine of vertebrates
-tubul	a little pipe	microtubules: hollow rods of protein found in the cytoplasm of almost all eukaryotic cells

▶

TABLE 3.1 Some common word parts used in biology, continued.

Word part	Meaning	Example
tunic	a covering	tunicates: members of the subphylum Urochordata, marine chordates without backbones; sedentary adults are enclosed in protective tuniclike coverings
turg-	swollen	turgor pressure: the force directed against a cell wall after an influx of water and the swelling of such a walled cell due to osmosis
tympan	a drum	tympanic membrane: another name for the eardrum
ultra-	beyond	ultracentrifuge: a machine that spins test tubes at the fastest speeds to separate liquids and particles of different densities
uni-	one	unisexual: referring to animals or plants with either male or female reproductive organs but not both
uro	tail	urochordates: marine chordates without backbones, commonly called tunicates; larval urochordates have tails
utric	a leather bag	utricle: a chamber behind the oval window in the inner ear that opens into the three semicircular canals
vacu-	empty	vacuole: sac that buds from the ER, Golgi, or plasma membrane in eukaryotic cells
vagin	a sheath	invagination: the infolding of cells
-valent	strength	covalent bond: an attraction between atoms that share one or more pairs of outer-shell electrons
vasa	vessel	vasocongestion: the filling of a tissue with blood caused by increased blood flow through the arteries of that tissue
vascula-	little vessel	vascular tissue: plant tissue consisting of cells joined into tubes that transport water and nutrients throughout the plant body
-vect	carried	convection: the mass movement of warmed air or liquid to or from the surface of a body or object
-ventriculo	ventricle, hollow organ	atrioventricular node: a region of specialized muscle tissue in the heart between the right atrium and right ventricle; it generates electrical impulses that primarily cause the ventricles to contract
vestigi-	trace	vestigial organs: structures of marginal, if any, importance to an organism; historical remnants of structures that had important functions in ancestors
villi	shaggy hair	microvilli: many fine, fingerlike projections of the epithelial cells in the lumen of the small intestine that increase its surface area
virul	poisonous	virulent virus: a virus that reproduces only by a lytic cycle, thus causing the death of the host cell
vitre	glass	vitreous humor: the clear, jellylike material that fills the posterior cavity of the vertebrate eye

TABLE 3.1 Some common word parts used in biology, continued.

Word part	Meaning	Example
vivi	alive	ovoviviparous: referring to a type of development in which young hatch from eggs that are retained in the mother's uterus
vora	eat	herbivore: a heterotrophic animal that eats plants
-xantho	yellow	xanthophyll: one of a class of pigment molecules that provide the yellow and brown colors to autumn leaves
xero-	dry	xerophytes: plants adapted to arid climates
xyl-	wood	xylem: the tube-shaped, nonliving portion of the vascular system in plants that carries water and minerals from the roots to the rest of the plant
-yl	substance or matter	adenylyl cyclase: an enzyme built into the plasma membrane of cells that converts ATP to cAMP
zea-	a grain	zeaxanthin: a blue-light photoreceptor involved in stomatal opening in plants; gives corn, saffron, and many other plants their characteristic yellow color
zoa, zoan	animal	bryozoan: colonial animals (phylum Ectoprocta) that superficially resemble mosses parazoan: sponges, which are animals lacking true tissues
zona	a belt	zona pellucida: the extracellular matrix of a mammalian egg

TIME TO TRY

Now you can practice putting your new knowledge to use. Match the following terms with their meanings.

1. _____ arthritis
2. _____ omnivore
3. _____ appendectomy
4. _____ pseudopodium
5. _____ thermocline

a) Narrow zone of temperature change in an ocean or lake

b) Surgical removal of the appendix

c) A cellular extension ("false foot") of amoeboid cells

d) Inflammation of the joints

e) An animal that eats both meat and plant material

Answers: 1. d. 2. e. 3. b. 4. c. 5. a.

How to Make a Jack-o'-Lantern: **Combining Word Roots**

When learning biological terms, you should know some simple rules. You may have noticed in **Table 3.1** that some roots end with "/o." That means the root can be used with or without an "o" added at the end. When the "o" is added, the modified root is referred to as the **combining form**. It is awkward to pronounce a word formed from a root that ends in a consonant letter and another root or suffix that starts with a consonant. For these roots, a vowel is inserted in the middle to make it easier to pronounce. This is most often an "o." For example, inside your body your abdominal cavity and your pelvic cavity are continuous, so they are often referred to as one cavity by combining the two roots: *abdomin* and *pelvic*. The resulting word would be *abdominpelvic*. Instead, we insert an "o" between the roots to get *abdominopelvic*, which rolls off the tongue more easily. This rule is the jack-o'-lantern construction.

TIME TO TRY

Combine the following word parts, then define the term.

1. gastr + -itis = ____________
   (stomach) (inflammation)
   What does this term mean? ____________
   ____

2. arthro + -pod = ____________
   (jointed) (foot)
   What does this term mean? ____________
   ____

3. hydro + -philic = ____________
   (water) (loving)
   What does this term mean? ____________
   ____

4. bi + -ennial = ____________
(two) (year)
What does this term mean? ____________
____

5. peri + -cardium = ____________
(around) (heart)
What does this term mean? ____________
____

Did you get those? The word parts in the first question combine to form *gastritis,* which means an inflammation or infection of the stomach. The pair in the next question form *arthropod,* the name for a large group of invertebrate animals with jointed appendages. The third pair combine to form *hydrophilic,* a term that describes substances that can mix with water. The fourth pair form *biennial,* a term that is used to describe plants that reproduce every two years. Finally, you combined the fifth pair to form *pericardium,* the name for the membrane that encloses the heart.

Here is another rule to remember—spelling counts! In some cases, changing a single letter can make a difference in the meaning of the word. For example, two terms that we use to discuss humans and their relatives and ancestors are *hominid* and *hominin.* Note that the only difference between the two is the last letter. However, hominid usually refers to the group of related primates including humans, chimpanzees, gorillas, and their extinct common ancestors, whereas hominin refers to only humans and their extinct ancestors. Only that one letter distinguishes these two terms. Here is another pair: *ilium* and *ileum.* The ilium is part of your hip bone, but the ileum is the last part of your small intestine.

Realize the importance of spelling—changing even one letter may change what you are referring to.

Another maneuver that can be tricky with these terms is converting them from the singular form to the plural form. Some of the rules are the same as in English. If you look at one muscle and then another muscle, you are indeed examining two muscle**s**, and more than one bone are bone**s**. But there are some unique rules for pluralizing many terms that you will encounter. More than one vertebra are vertebra**e**, and the small bone at the end of each of your fingers is a phalanx, but all of these bones in your fingers together are called phalang**es**. **Table 3.2** shows the rules to guide you with pluralization.

TABLE 3.2 Basic rules for changing words in singular form to plural.

If the word ends in	Do this first	Then add	Examples
-a		Add *-e*	*larva* becomes *larvae*
-ax	Drop *-ax*	Add *-aces*	*thorax* becomes *thoraces*
-ex or *-ix*	Drop *-ex* or *-ix*	Add *-ices*	*cortex* becomes *cortices*
-ma		Add *-ta*	*stoma* becomes *stomata*
-is	Drop *-is*	Add *-es*	*anastomosis* becomes *anastomoses*
-nx	Change *-x* to *-g*	Add *-es*	*larynx* becomes *larynges*
-on	Drop *-on*	Add *-ia*	*ganglion* becomes *ganglia*
-us	Drop *-us*	Add *-i*	*nucleus* becomes *nuclei*
-um	Drop *-um*	Add *-a*	*ischium* becomes *ischia*
-y	Drop *-y*	Add *-ies*	*biopsy* becomes *biopsies*

In this chapter, we threw a lot of new terms at you. Most chapters in your textbook will do the same. At times, you may think that learning all the terms is impossible. Just when you think you know the name of a structure, you might learn that it has another name as well. Think about the red blood cell that carries oxygen in your blood. Do you know any other names for a red blood cell? You might—it is also called an **erythrocyte**, or it goes by the abbreviation RBC.

Multiple names are not uncommon in biology. In fact, you might hear your instructor say "Why name it once if you can name it two or three times and confuse people?" At times, the confusion may seem intentional, but realize that there are many ways to name things—formal names (erythrocyte) and informal names (red blood cell), for example. Mastering the language can be challenging, but you are now armed with important tools to help you. You will also find that many terms are defined in your textbook, either in the chapters or in the glossary. And you can always purchase a biological dictionary or refer to an online version—they are invaluable.

But remember the traveler who helped open our chapter. You now know to start your journey through each chapter in your textbook by learning the language. Know the terms so you can understand the material. Don't just memorize words—look up their meanings, look at how they are built, make and use flashcards, and be able to use your new words. Get in the habit of talking biology with your study group and with other people in your life, if possible. Learn what you can now and the rest will come as you go.

People often move to foreign-speaking countries and learn the language as they go. They learn by experience and practice. Students have been surviving in the foreign-sounding world of biology for ages as well, and you have already had a crash course in the language. Enjoy your adventure!

Final Stretch!

Now that you have finished reading this chapter, it is time to stretch your brain a bit and check how much you learned.

RUNNING WORDS

At the end of each chapter, be sure you have learned the language. Here are the terms introduced in this chapter with which you should be familiar. Write them in a notebook and define them in your own words, then go back through the chapter to check your meaning, correcting as needed. Also try to list examples when appropriate.

Descriptive term	Prefix
Eponym	Root
Abbreviation	Suffix
Acronym	Combining form

WHAT DID YOU LEARN?

Try these exercises from memory first, then go back and check your answers, looking up any items that you want to review. Answers to these questions are at the end of the book.

PART A: USING TABLE 3.1, MATCH THE FOLLOWING TERMS WITH THEIR DESCRIPTIONS.

1. _____ leukocyte
2. _____ endocarditis
3. _____ pinocytosis
4. _____ acromegaly
5. _____ cytoplasm
6. _____ mammography
7. _____ photosynthesis
8. _____ dendrochronology
9. _____ gastroenteritis
10. _____ quadriplegia

a) Paralysis of all four limbs
b) The process of converting light energy to chemical energy
c) Inflammation of the stomach and intestine
d) The study of tree rings to determine the age of wood
e) A white blood cell
f) Inflammation of the lining inside the heart
g) The fluid and structures inside a cell
h) An imaging technique used to assess the breasts
i) The process of cells taking in droplets of liquid
j) A condition characterized by chronic growth of the tips of bones

PART B: ANSWER THESE QUESTIONS.

1. Which of the following is an eponym?

   ____

   a) CPR b) SIDS
   c) cardiac sphincter d) Bartholin's glands

2. Which of the following is an acronym?

   ____

   a) CPR b) SIDS
   c) cardiac sphincter d) Bartholin's glands

3. Using Table 3.1, construct a descriptive term for each of the following:

   Splitting of a glucose molecule __________

   Process of moving materials out of a cell

   ____

   Plant that grows on top of another plant

   ____

4. Using the same table, define each of the following terms:

   Cytokinesis ______________________________

   ____

   Chromosome ______________________________

   ____

   Thigmotropism ______________________________

   ____

5. Provide the plural form of the following terms:

   pharynx ______________________________

   mitochondrion (structure inside a cell)

   ____

   coxa (hip bone) ______________________________

4 Biology Basics

How Life Works

When you complete this chapter, you should be able to:

- Understand the biological hierarchy of organization.
- Understand basic principles of biology.
- Discuss the major groups of living organisms.
- Understand the basic process of scientific inquiry.

Your Starting Point

Answer the following questions to assess your knowledge.

1. The scientific study of the interactions between organisms and their environment is called ______________________.
2. What is meant by "Form fits function?" ______________________
____
3. Where do plant cells get the energy they use for work? ______________________
____
4. How many major categories of bacteria are there? ______________________
5. What process results in the evolution of adaptive traits in groups of organisms? ______________________
____

Climbing the Ladder: **The Biological Hierarchy of Organization**

Now that you are armed with some new language skills, let's peek at what lies ahead in your study of biology. You have a lot to learn along the way, but your journey is also one of self-discovery—you are learning about all forms of life, including yourself. That should make your adventures in biology quite exciting! This chapter will introduce some basic concepts and principles to guide your path. Let's begin with how your course will be organized.

Answers: 1. Ecology. 2. A part's structure reflects the job it does. 3. From the sun. 4. Two. 5. Natural selection.

In science, we like to categorize. One classification system is known as the biological hierarchy of organization. This hierarchy begins at the simplest level of structural organization and ends at the most complex, as shown in **Figure 4.1** on the next page. Your textbook and your course will probably follow this pattern. These levels, from the simplest to the most complex, are the atom, molecule, macromolecule, organelle, cell, tissue, organ, organ system, organism, population, community, and ecosystem. The largest and most complex level of all is the biosphere, which encompasses the environments that are inhabited by life.

TIME TO TRY

Examine Figure 4.1. Note that the bottom three levels (atom, molecule, and macromolecule) are part of the science of chemistry. The next two levels—organelle and cell—are covered by another science discipline. Look back at Table 3.1 and determine the name of this science that studies cells and their structures.

What is it? ______________________________

The most complex level is the ecosystem. Again, refer to Chapter 3 and determine the name of this branch of science.

What is it? ______________________________

The bottom three levels shown in Figure 4.1 are **chemistry**, the study of elementary forms of matter, which we will review in Chapter 5. The next level is cytology or cell biology, which we will explore in Chapter 6. **Cell biology** is the study of the structure and function of cells, but there are several related biological disciplines at this level as well, including **microbiology**, the study of microscopic organisms such as bacteria and viruses, and **molecular genetics**, the study of the genetic material. The next more complex levels—organ, organ system, and organism—are the realm of **anatomy and physiology**. The next level, that of populations, is covered by **population genetics**, the study of inheritance between generations of organisms, as well as by **evolutionary biology**, the study of populations and species through time. **Paleontology** also covers this level, as it is the study of extinct species. Finally, the most

FIGURE 4.1 The biological hierarchy of organization.

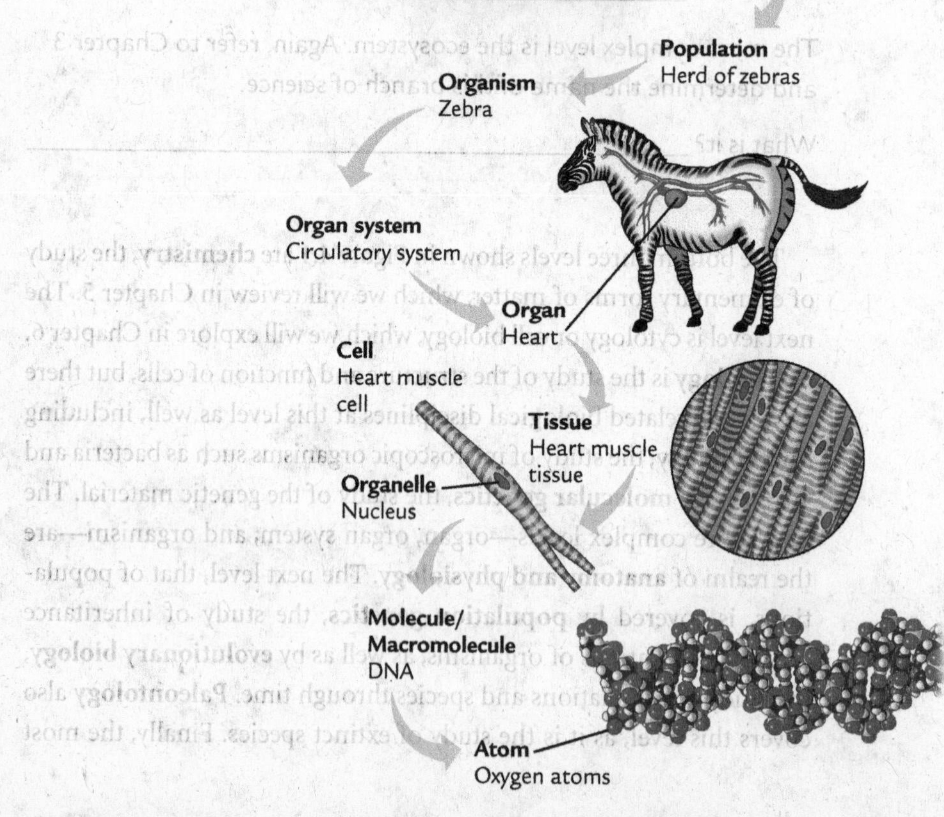

complex level—populations, the community, ecosystem, and biosphere—are studied in **ecology**. There are many other subdisciplines in biology that we haven't mentioned here.

You can see from this brief overview the enormous scope of biology. Try thinking of this scope as having two dimensions. The "vertical" dimension, which is shown in Figure 4.1, is the size scale that reaches all the way from the microscopic molecules and cells that make up organisms to the global scale of the entire living planet. But you can also see from this list that biology's scope has a "horizontal" dimension stretching across the great diversity of species, now and over life's long history. Next, let's take a look at each of the levels of the hierarchical organization by size in a bit more detail.

All matter is made of **chemical elements**. The smallest piece of an element is an **atom**, and atoms can unite to form **molecules**. To understand this, consider water, possibly life's most important molecule. One atom of the element oxygen combines with two atoms of hydrogen to form a molecule of water (H_2O). Water molecules are quite small, but some molecules, such as fat, DNA, starch, and proteins, are rather large, so they are referred to as **macromolecules**. Atoms, molecules, and macromolecules provide the nutrients and building materials living organisms need to stay alive and healthy, and they participate in chemical reactions that do all of the work performed by those organisms.

Macromolecules can unite to form complex structures called **organelles** that carry out functions inside cells. Organelles include cellular structures like the nucleus or a mitochondrion. **Cells** contain the combination of organelles necessary to sustain life, so cells are the first level of organization that we consider to be alive. In fact, cells are the basic unit of all living organisms. **Tissues** are groups of cells organized to perform some common function. For example, muscle tissue contracts to provide movement. Tissues then organize into larger functional units called **organs**, such as the lungs, heart, and liver. Each organ performs at least one specific job. Multiple organs combine to form **organ systems**, each with some overall function. For example, the organs in the cardiovascular system are the heart and the blood vessels. The heart is the pump and the vessels are the pipes, so to speak, through which the blood travels to deliver nutrients and oxygen to the cells and to haul away their wastes. Collectively, the organ systems do the work needed to keep the **organism** alive.

PICTURE THIS

You decide to build a house. You use lumber that came from trees, which were once living organisms. Explain how the first five levels of the biological hierarchy apply to the tree from which your lumber came. ____

____

____

____

Trees are made of cells containing organelles that allow the tree to grow and form woody material, such as cellulose, which is a macromolecule. Macromolecules are built from smaller molecules and they, in turn, are built from atoms.

In the biological hierarchy of organization, there are levels beyond the individual organism. A **population** is a localized group of organisms belonging to the same species; populations of different species living in the same area make up a biological **community**. Interactions between the living organisms in a community and the nonliving features of the environment, such as sunlight and water, form an **ecosystem**.

✔ QUICK CHECK

Fill in the missing levels from the biological hierarchy of organization.

1. atom, molecule, ______________, organelle
2. cell, ______________, organ, ______________, organism,
3. ______________, community, ecosystem.

Answers: 1. macromolecule. 2. tissue; organ system. 3. population.

"The whole is greater that the sum of its parts." This familiar adage captures the important concept that a combination of components can form a more complex organization called a **system**. Examples of biological systems are cells, organisms, and ecosystems. To understand how such systems work, it is not enough to have a "parts list," even a complete one. Keep in mind as you read the following basic biological rules that all the parts of living organisms are themselves biological systems and are at the same time part of larger systems with unique properties that emerge as the complexity of interactions between the parts increases.

Thinking about Life: **Basic Biological Principles**

Most of biology is beautifully simple and understandable, as you will soon see. At first you may be tempted to just memorize names, but you must move beyond memorization to truly understand the processes involved in the biology. When studying the course material, always look for the logic and reasoning behind it. Your task will be easier if you understand some basic principles that govern life. These principles will help you understand why living organisms do what they do. These concepts should also help guide your thinking as you try to understand the "whys" behind your learning.

LIFE BEGINS AT THE CELL

The cell has a unique place in the biological hierarchy of organization. It is the lowest level of structure that has the ability to perform all of the activities necessary for life, including that of reproduction.

All organisms are composed of cells. Some organisms are made up of only a single cell; these are called **unicellular** organisms, and most of them are visible only through a microscope (hence the name microscopic organisms). In addition, cells form the subunits that make up the tissues and organs of plants, animals, and other organisms composed of more than one cell, called **multicellular** organisms. Whether an organism is unicellular or multicellular, the cell is still the organism's basic unit of structure and function. The ability of cells to divide to form new cells is the basis for all reproduction and for the growth and repair of multicellular organisms, including humans.

All cells have some things in common. Most importantly, all cells use **DNA** as the chemical material of **genes**, the units of inheritance that transmit information from parents to offspring. Cells also show many differences, depending upon their functions. There are two main types of cells, based on the organization of their internal structure. All cells are either **prokaryotic** or **eukaryotic**. *Pro-* means "before" and *karyo-* means "nucleus," so the term prokaryote means "before the nucleus." Prokaryotic cells lack internal membranes so, indeed, they don't have an enclosed nucleus. Prokaryotes are unicellular and microscopic. Although these cells may appear simple, they are quite successful and exist all around us—all bacteria are prokaryotic. The term eukaryote, in contrast, comes from Latin: *eu-* means "true," and these cells have a true nucleus. The nucleus encloses the cell's DNA. Eukaryotic cells also have other internal compartments called organelles. You will learn more about cell structure and function in Chapter 6.

FORM FITS FUNCTION

A common theme throughout biology is that "Form fits function." This means that all the parts of an organism have specific structures that allow them to perform their jobs most efficiently. A part's shape and organization reflect what it does, and similarly the job that is needed affects the structure that a part will have.

Let's consider the human heart (**Figure 4.2**). The heart's job is to circulate the blood, and it does this by constantly pumping blood along its way. To accomplish this, the heart has receiving chambers, the atria, and sending chambers, the ventricles. Some of the blood returning to the heart comes from the lungs carrying a rich supply of oxygen, and is ready to head out through the body to supply all the cells. The rest of the blood returning to the heart has just been to the cells and dropped off much of its oxygen for their use, so it needs to go to the lungs to get more. Your heart is divided into right and left sides. This allows your blood to travel through two different paths—the right side sends the deoxygenated blood out to the lungs, and the left side sends the freshly oxygenated blood out to your body's cells.

Your blood circulates best if it is always moving in a single direction—forward. Doorways called valves are part of the heart's anatomy. One set allows blood to enter the ventricles and the other set allows it to leave

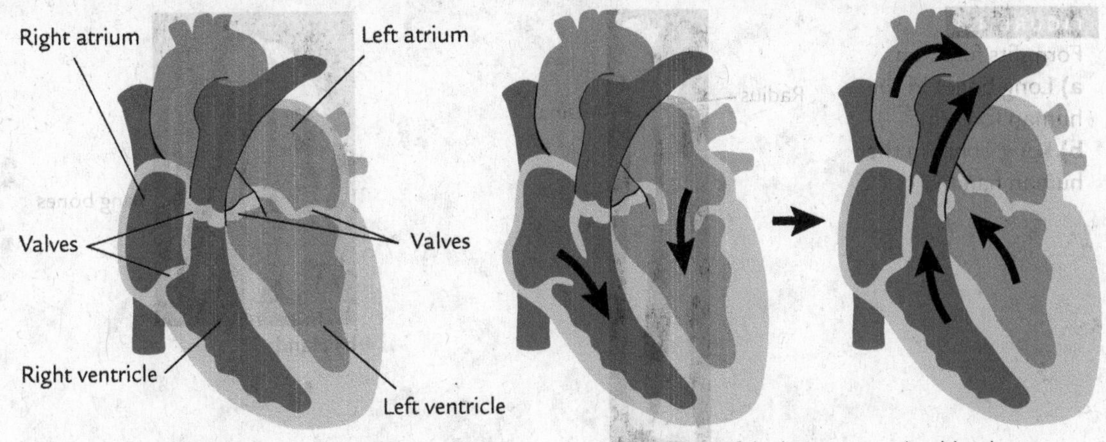

FIGURE 4.2 The anatomy of the human heart. **a)** The heart chambers and valves. **b)** As the ventricles fill, valves at the entrances into the ventricles are open but valves at the exits are closed. As the ventricles empty, valves at their entrances are closed but valves at their exits are open.

these chambers. To fill the ventricles, the doors into the ventricles are open but the exit doors are closed (**Figure 4.2b**). When the ventricles contract, the entrance doors close so the blood must move forward, pushing the exit doors open so it can leave. Clearly the heart's anatomy is uniquely suited to its function. In other words, the heart's form fits its function.

TIME TO TRY

Examine **Figure 4.3** on the following page to answer the following question.

1. Figures 4.3 a) and b) both show long bones, but the long bones in your hand are much shorter than the long bones of your forearm. How does the function of these two areas explain this structural difference in their long bones?

____

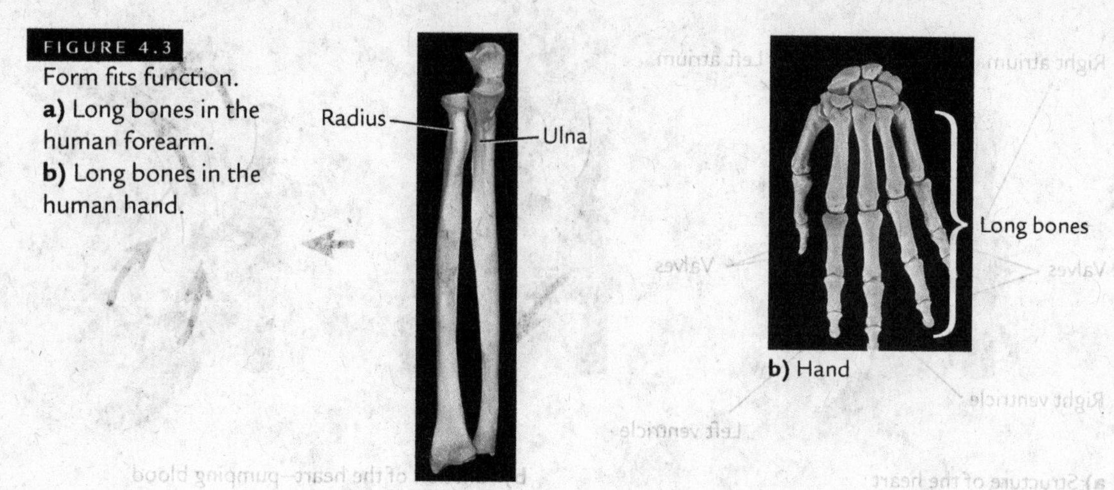

FIGURE 4.3
Form fits function.
a) Long bones in the human forearm.
b) Long bones in the human hand.

Your forearm contains only two long bones, but there are 19 much shorter "long" bones in your hand. This allows the hand to move into many different shapes and to grip objects tightly, making it one of our most versatile tools.

So when you examine a structure, the way it is built should give you clues about what it does. Similarly, if you know what a structure does, you should be better able to understand how and why it is built the way it is.

✔ **QUICK CHECK**

What is meant by "Form fits function?" ______________________

____

Answer: A part is organized in a manner that makes it most suited to the job it needs to perform. Its structure reflects its function.

The structure of the parts of an organism reflects their function. ■

REGULATION AND HOMEOSTASIS

To maintain the utmost efficiency, living organisms must exist in an optimal working environment. To achieve this, they have numerous regulatory mechanisms, most of which require energy, that work to maintain a relatively constant internal environment. This internal constancy is called **homeostasis.** You can think of it as maintaining the right balance of conditions within an organism.

TIME TO TRY

Turn back to Table 3.1 to determine the literal meaning of the word *homeostasis*. ______________________________

Homeostasis literally means "to stay the same." Most work done within an organism is the result of chemical reactions. These reactions occur most efficiently if there is a relatively constant temperature, the right amount of water, and the right amount of chemicals, for example. We have an acceptable normal range of values for each of these. We know normal human body temperature is around 98.6°F (37° Celsius). If our body cools too much, chemical reactions will occur more slowly or not at all. If our temperature gets too hot, chemical reactions speed up and some chemicals may be destroyed.

PICTURE THIS

Imagine you have some prized plants growing in your garden. Before leaving for two weeks during a brutal summer drought, you set a timer on your garden hose that will let it run at just the right flow for 2 hours each day while you are gone. You return home to find that your once-gorgeous flowers are now dried brown sticks! Some investigation reveals that although your water did run for 2 hours each day, the flow was reduced because some pesky varmint ate a hole in your special hose!

1. Why did your plants die? ______________________________

____

▶

2. In the body, water is a critical substance that also carries other nutrients and waste products to and from your cells. What would happen to this transport if you had too little water in your body?

____

3. How might that affect your cells? ______________

____

4. How would overwatering affect your plants? ______________

____

5. How might overwatering affect your cells? ______________

____

Living organisms are composed mostly of water. For example, it is the main component of human cells and blood. Water balance is absolutely critical in the human body. If you have too little water, nutrients cannot be adequately transported to your cells and wastes can accumulate to toxic levels. Your cells will work less efficiently and may die. In plants, too much water dilutes the nutrients the plant cells need and causes them to swell and perhaps die. In the human body, too much water has the same effect, and when brain cells swell, death can occur.

Let's consider another example. Think about how your home's furnace and air conditioner work. You set the desired temperature at, let's say, 70°F. If the air gets colder than that, your furnace turns on and heats the air until it is back to 70°, then the furnace shuts off. If the air temperature exceeds 70°, your air conditioner turns on and cools the air until it reaches 70° again, then it turns off. Notice that there are two possible problems—the temperature can go too low or it can climb too high. Both problems have a solution, but because the problems are opposites, so are the solutions—one raises the temperature but the other lowers it. Both solutions stop once the problem is corrected—they are self-limiting. This kind of control is called **negative feedback**, and it is the most common control mechanism used in animal physiology and the main way your body maintains homeostasis.

WHY SHOULD I CARE?

Homeostasis and negative feedback are recurring themes in biology. The normal state of an organism provides optimal health. Any significant fluctuation away from the normal range can quickly impair normal function and perhaps become critical. These fluctuations form the basis of disease diagnosis. Negative feedback is the control system by which animal bodies correct these errors and restore health.

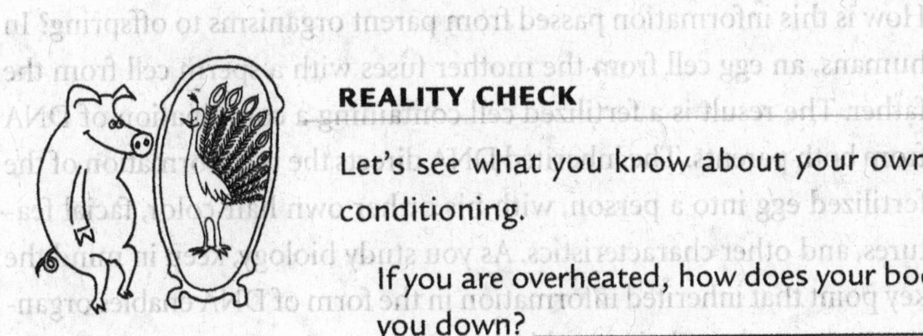

REALITY CHECK

Let's see what you know about your own body's furnace and air conditioning.

If you are overheated, how does your body respond to try to cool you down? ______________________

If you are too cold, how does your body respond to try to warm you up? ______________________

If you are too hot, you sweat more and the blood vessels in your skin dilate to bring the overheated blood to the surface of the skin where heat can be more easily lost. If you are too cold, you sweat less and blood stays away from the surface of the skin to keep your internal organs warmer. You shiver, which involves rapid contraction of your muscles. That activity generates heat.

Balance, or homeostasis, is important in many aspects of biology. Keep this concept in mind and you will be able to more easily predict how organisms might respond in different circumstances.

REPRODUCTION AND INHERITANCE

It is only through reproduction that organisms pass on their genes and the species survives. In fact, many biologists maintain that the ultimate purpose of our existence is reproduction.

And tied to reproduction is **inheritance**, the sum of the genetic characters that are passed from parents to offspring. You've noticed that children tend to look like their parents. This happens because the genes for certain characteristics are inherited by the children from their parents. Genes are responsible for family inheritance.

Recall that genes are made of DNA, which stores information. Each cell in your body contains a copy of all the DNA that you inherited from your mother and father. Before a cell divides, it copies its DNA and passes this genetic information on to each of the two cells it produces. How is this information passed from parent organisms to offspring? In humans, an egg cell from the mother fuses with a sperm cell from the father. The result is a fertilized cell containing a combination of DNA from both parents. The inherited DNA directs the transformation of the fertilized egg into a person, with his or her own hair color, facial features, and other characteristics. As you study biology, keep in mind the key point that inherited information in the form of DNA enables organisms to reproduce their own kind.

Reproduction enables survival of the species. During reproduction, genes are passed from parent to offspring in the process of inheritance. ■

✔ **QUICK CHECK**

1. What is homeostasis? ______________________

____

2. What is the most common way to regulate homeostasis?

____

3. Biologically speaking, what is the goal of our existence?

____

Answers: 1. Maintenance of a relatively constant internal environment. 2. Negative feedback. 3. Survival for the purpose of reproduction.

INTERACTION WITH THE ENVIRONMENT

Life doesn't exist in a vacuum. Each organism interacts continuously with its environment, which includes other organisms as well as nonliving components of the environment. The roots of a tree, for example, absorb water and minerals from the soil. Leaves take in a gas called carbon dioxide from the air. The tree also releases oxygen to the air, and its roots help form soil by breaking up rocks. Both organism and environment are affected by the interactions between them. The tree also interacts with other living things, including microscopic organisms in the soil that are associated with the plant's roots and animals that eat its leaves and fruit. We are, of course, among those animals. As you learned, ecology is the branch of biology that investigates these relationships between organisms and their environments, at the level of the ecosystem.

LIFE REQUIRES ENERGY

All living organisms use energy. **Energy** is defined as the ability to do work. We can look at energy use on at least two levels, those of the cell and the ecosystem. Energy is used by all cells to allow them to move, grow, reproduce, and perform all the other activities required for life. Together, all of the chemical reactions that take place in a cell are called its **metabolism**. Cells that can harness the power of the sun are those that can perform **photosynthesis**, in which solar energy is converted to the chemical energy contained in sugars. Those cells that cannot perform photosynthesis use chemical energy obtained from photosynthetic cells.

Therefore, most ecosystems are run on solar power converted to chemical energy by plants and other photosynthetic organisms. (**Figure 4.4** depicts this energy flow through an ecosystem.) Such organisms are called producers. The energy produced then flows to consumers, which are organisms, such as animals, that feed on plants either directly (by eating plants) or indirectly (by eating animals that eat plants). Thus energy in an ecosystem flows from sunlight to producers and then to consumers. In the process some energy is lost as heat. At both the cellular and ecosystem levels, energy is necessary for life.

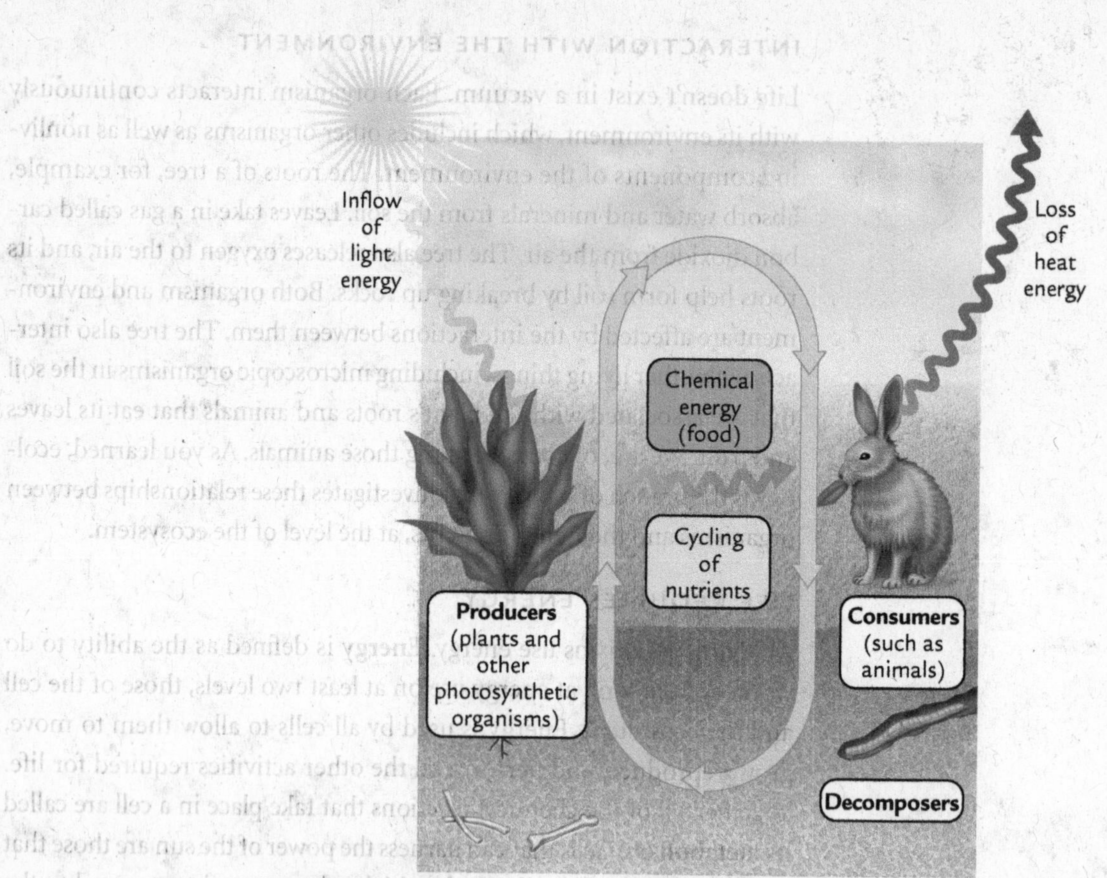

FIGURE 4.4 Energy and nutrient flow in an ecosystem. Living is work, and work requires that organisms obtain and use energy. Most ecosystems are solar powered. The energy that enters an ecosystem as sunlight exits as heat, which all organisms dissipate to their surroundings whenever they perform work. In contrast, the nutrients within an ecosystem are recycled.

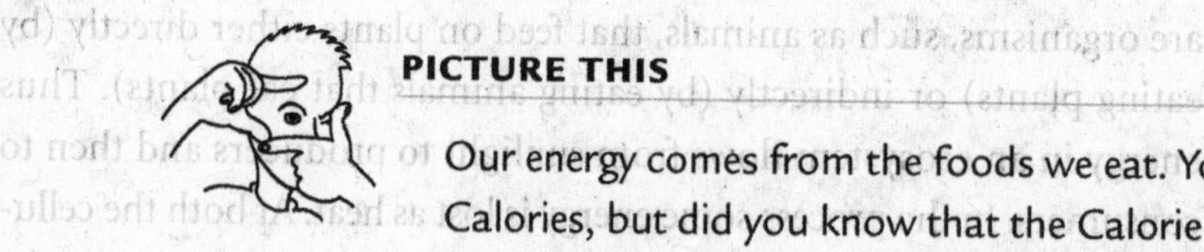

PICTURE THIS

Our energy comes from the foods we eat. You know that foods have Calories, but did you know that the Calorie is actually the basic measuring unit for energy? Think about our energy, or Calories, as being the money of life. To a certain extent, we can control our energy "income."

1. How? ______________________________

2. If we do not take in enough energy "money," we cannot meet all of our energy demands, or, in effect, we cannot pay our bills. What happens in your home if you don't pay your power or your water bill? ______________________________

3. What would happen to your cells if you did not give them enough energy? ______________________________

4. If we earn more real money than we spend, what do we wisely do with the extra? ______________________________

5. What do you think the body does with extra energy "money" it takes in? ______________________________

Most extra energy, again meaning Calories, is stored in the body as fat, regardless of what we eat. This is the body's version of saving for a rainy day, and we gain weight. As with our bank accounts, to lose stored income (weight), we need to spend more than we are taking in. That may be, unfortunately, very easy to do with real money, but your body "bank" doesn't make it so easy to take those extra Calories out of storage because it, like all organisms, conserves the energy that is critical to life.

Obtaining all the materials needed to make and maintain an organism's parts requires considerable energy. To minimize this cost, all organisms recycle. Many molecules are broken down to atoms and then built back into other molecules, over and over again. Think about a child's building block set. Children can spend hours building things, taking them apart, then building new ones—a car is stripped down, then becomes a castle, which is stripped down and rebuilt into a robot. Like the child, all living organisms reuse both energy and materials.

All living organisms require energy to do work. ■

ADAPTATION AND EVOLUTION

Can you find the three animals in **Figure 4.5**? These organisms are three species of insects called mantids. Their shapes and colors enable them

a) A Trinidad tree mantid that mimics dead leaves

b) A leaf mantid in Costa Rica

c) A flower mantid in Malaysia

FIGURE 4.5 Camouflage as an example of evolutionary adaptation. Related species of insects called mantids have diverse shapes and colors that evolved in different environments.

to blend into their backgrounds. This camouflage makes the mantids less visible to animals that feed on insects. It also makes them less visible to the insects the mantids feed on!

The unique characteristics that camouflage each mantid species are examples of adaptations. An **adaptation** is an inherited trait that helps the organism's ability to survive and reproduce in its particular environment.

How do mantids and other organisms adapt to their environments? Part of the answer is the variation among individuals in a population. As you learned earlier, a population is a localized group of organisms belonging to the same species. Just as you and your classmates are not exactly alike, individuals of all populations, including mantids, also vary in some of their traits. These variations reflect each individual's particular combination of inherited genes. And this hereditary variation is the raw material that makes it possible for a population to adapt to its environment. If a particular variation is helpful, individuals with the variation may live longer and produce more offspring than those that do not have it. This process is called **natural selection** because it works by the natural environment "selecting" certain inherited traits.

Figure 4.6 illustrates a hypothetical example of natural selection in a beetle population. The individual beetles vary in their coloring, from light gray to charcoal. Each beetle's color is determined by its genes. Now suppose that the soil has recently been blackened by a fire. For birds that eat the beetles, it is easiest to spot the beetles that are lightest in color. On average, the darker beetles have a better chance of surviving and reproducing, passing their genes for dark color on to their offspring. In contrast, the lighter beetles are captured more easily, and fewer survive to produce offspring. After many generations, most of the beetles in the population are dark. This abundance of dark color is an adaptation of the beetle population to its environment.

FIGURE 4.6 Natural selection. In this hypothetical example, darker beetles are more likely to survive longer and reproduce, passing their genes on to offspring. Over time, this changes the abundance of dark- and light-colored beetles in the population.

1) A beetle population includes individuals of different colors.

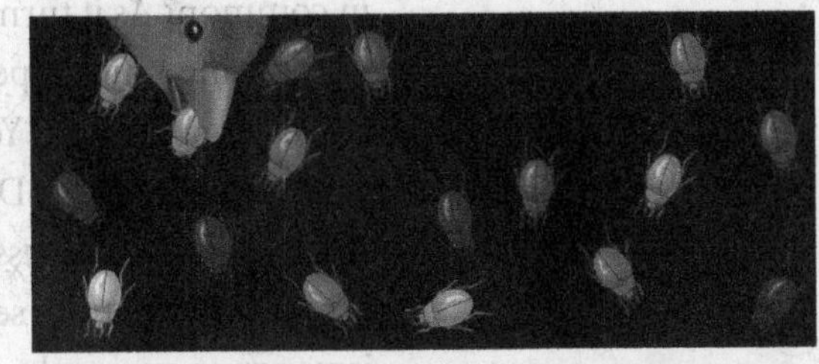

2) Birds capture more light beetles than dark beetles.

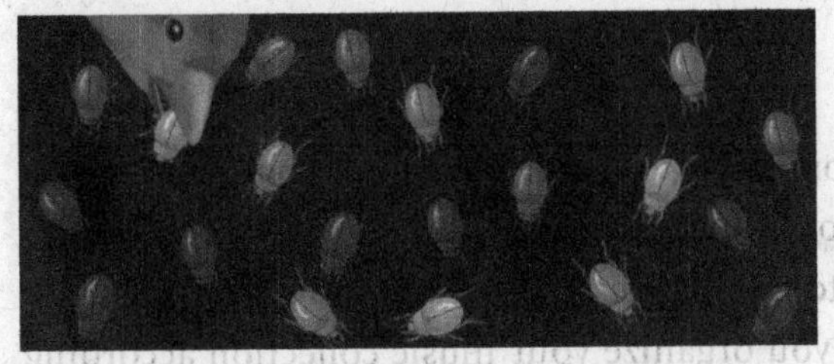

3) Survivors (mostly dark beetles) reproduce.

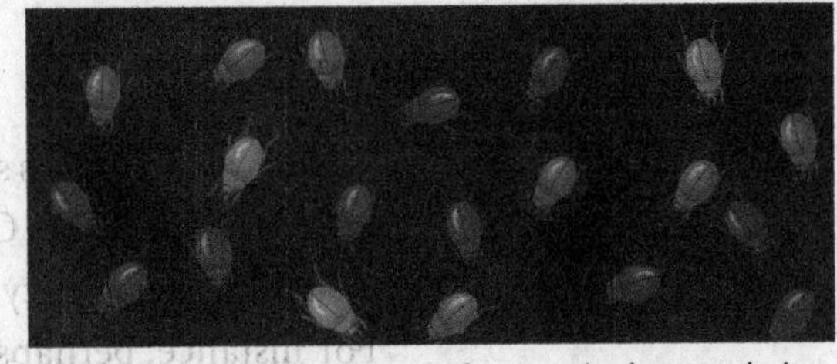

4) Dark beetles become more frequent in the population over time.

Natural selection is the mechanism by which evolution occurs. The word *evolution* means "a process of change." Biologists use the term **evolution** specifically to mean a generation-to-generation change in the proportion of different inherited genes in a population. For example, in the beetle example, genes for dark color are becoming more common and genes for light color are becoming less common over the generations of beetles. The beetle population is said to be undergoing evolution, or evolving. If this process continues, the dark beetle population may change enough to be considered a new species.

UNITY IN DIVERSITY

Diversity is a hallmark of life. Biologists so far have identified and named about 1.8 million species. This **biodiversity** includes over 290,000 types of plants, 52,000 kinds of vertebrates (animals with backbones), and 1,000,000 kinds of insects (more than half of all known forms of life). Biologists add thousands of newly identified species to the list each year. Estimates of the total diversity of species range from about 10 million to over 200 million. Whatever the actual number, the vast variety of life widens biology's scope.

If life is so diverse, how can biology have unifying basic principles? What, for instance, can a tree, a mushroom, and a human possibly have in common? As it turns out, a great deal! Underlying the diversity of life is a striking unity, especially at the lower levels of the biological hierarchy of organization. You have already seen one example: the universal genetic language of DNA. That fundamental fact connects all of life. Another is the process that has brought about this amazing diversity: evolution by natural selection. Above the cellular level, however, organisms are so variously adapted to their ways of life that describing and classifying biological diversity remains an important goal of biology.

Classification: **Grouping Life**

Biological diversity is something to relish and preserve, but it can also be a bit overwhelming. Confronted with such complexity, there seems to be a human tendency to group diverse items according to similarities. For instance, perhaps you organize your music collection according to artist. And then maybe you group the various artists into broader categories, such as dance music, party music, exercise music, and study-time music. In the same way, grouping species that are similar is natural for us. We may speak of squirrels and butterflies, though we recognize that many different species belong to each group. We may even sort groups into broader categories, such as rodents (which include squirrels) and insects (which include butterflies). **Taxonomy**, the branch of biology that names and classifies species, formalizes this ordering of species into a series of groups of increasing breadth. You will learn more about these in your biology course. For now, we will focus on kingdoms and domains, the broadest units of classification.

Until the last decade, most biologists divided the diversity of life into five main groups, or kingdoms. (The most familiar two are the plant and animal kingdoms.) But new methods, such as comparisons of DNA among organisms, have led to an ongoing reevaluation of the number and boundaries of kingdoms. Various classification schemes are now based on six, eight, or even more kingdoms. But as the debate continues on the kingdom level, there is broader consensus that the kingdoms of life can now be grouped into three even higher levels of classification called domains. The three domains are named Bacteria, Archaea, and Eukarya (**Figure 4.7**).

FIGURE 4.7
The three domains of life.

Domain Bacteria

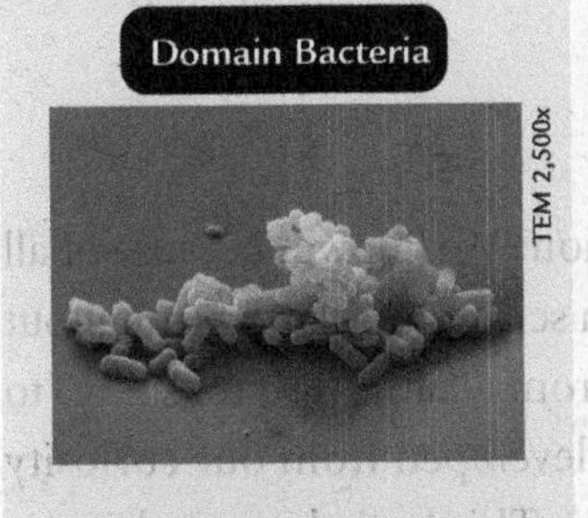

Domain Archaea

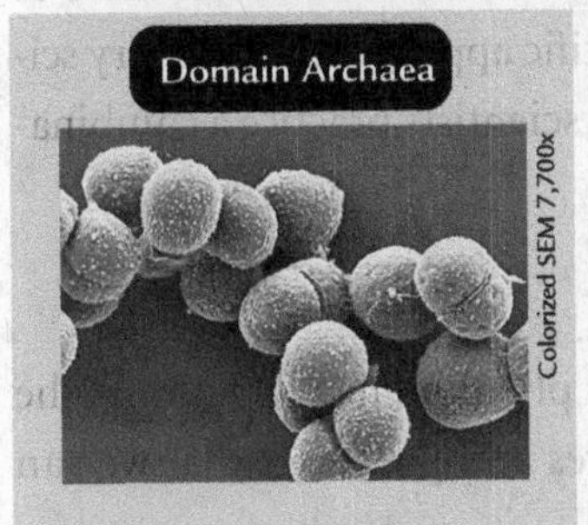

Domain Eukarya

Kingdom Plantae

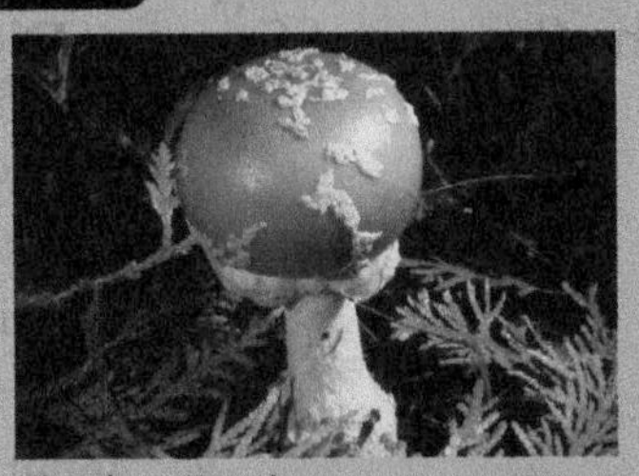

Kingdom Fungi

Kingdom Animalia

Protists (multiple kingdoms)

The first two domains, **domain Bacteria** and **domain Archaea**, identify two very different groups of organisms that have prokaryotic cells. In the five-kingdom system, bacteria and archaea were combined into a single kingdom. But newer evidence suggests that the organisms known as archaea are at least as closely related to eukaryotes as they are to bacteria.

All the eukaryotes (organisms with eukaryotic cells) are grouped into at least four kingdoms (depending on the classification scheme) in the **domain Eukarya.** Kingdom Protista consists of eukaryotic organisms that are generally single-celled—for example, the microscopic protozoa, such as the amoebas. Many biologists extend the boundaries of the kingdom Protista to include certain multicellular forms, such as seaweeds, that seem to be closely related to the unicellular protists. (Other biologists split the protists into multiple kingdoms.) The remaining three kingdoms—Plantae, Fungi, and Animalia—consist of multicellular eukaryotes. These three kingdoms are distinguished partly by their modes of nutrition. As you learned, plants are producers that make their own food by using the energy of sunlight. Fungi and animals are both consumers. Fungi are mostly decomposers that absorb nutrients by breaking down dead organisms and organic wastes, such as leaf litter and animal feces. Animals obtain food by ingestion, which is the eating and digestion of other organisms. It is, of course, the kingdom to which we belong.

The Process of Science: **Discovery and Hypothesis**

But how do we know all of this information? We know it because of all of the people who have been interested in science now and throughout our history. The word *science* is derived from a Latin verb meaning "to know:" Science is a way of knowing. It developed from our curiosity about ourselves and the world around us. This basic human drive to understand is manifest in two main scientific approaches: discovery science and hypothesis-driven science. Most scientists practice a combination of these two forms of inquiry.

DISCOVERY SCIENCE

Science seeks natural causes for natural phenomena. This limits the scope of science to the study of structures and processes that we can observe and measure, either directly or indirectly with the help of tools, such as microscopes, that extend our senses. This dependence on observations that other people can confirm demystifies nature and distinguishes science from belief in the supernatural.

Verifiable observations and measurements are the data of discovery science. In our quest to describe nature accurately, we discover its structure. In biology, discovery science enables us to describe life at its many levels, from ecosystems down to cells and molecules. A recent example of this type of science is the sequencing of the human genome, which is all of our DNA; the work that produced this information wasn't a set of experiments, but a detailed dissection and description of our genetic material.

Discovery science can lead to important conclusions based on a type of logic called inductive reasoning. An inductive conclusion is a generalization that summarizes many concurrent observations. "All organisms are made of cells" is an example. That induction was based on two centuries of biologists discovering cells in every biological specimen they observed with microscopes. The careful observations of discovery science and the inductive conclusions they produce are fundamental to our understanding of nature. They are often the basis for scientific **theories**, a term that is used to mean comprehensive explanations supported by abundant evidence (in contrast to how the word *theory* is used in our everyday lives to imply that an idea is speculative).

HYPOTHESIS-DRIVEN SCIENCE

The observations of discovery science engage inquiring minds to ask questions and seek explanations. Ideally, such investigation consists of what is called the **scientific method**. This method involves the following series of six steps:

1. Make an observation,
2. ask a question,
3. suggest an explanation or hypothesis,
4. make a prediction,

5. devise and perform a test, and

6. possibly rework the hypothesis and prediction depending on the results of the test.

The key element of the method that drives most modern science is called hypothetico-deductive reasoning or, more simply, hypothesis-driven science.

A hypothesis is a tentative answer to some question—an explanation on trial. It is usually an educated guess. We all use hypotheses in solving everyday problems. Let's say that your flashlight fails during a camping trip (**Figure 4.8**). That's an observation. The question is obvious: Why doesn't the flashlight work? A reasonable hypothesis based on past experience is that the batteries in the flashlight are dead.

The *deductive* in hypothetico-deductive reasoning refers to the use of deductive logic to test hypotheses. Deduction contrasts with induction, which, remember, is reasoning from a set of specific observations to reach a general conclusion. In deduction, the reasoning flows in the opposite direction, from the general to the specific. From general premises, we extrapolate to the specific results we should expect if the premises are true. If all organisms are made of cells (premise 1), and humans are organisms (premise 2), then humans are composed of cells (deductive prediction about a specific case).

In the process of science, the deduction usually takes the form of predictions about what outcomes of experiments or observations we should expect *if* a particular hypothesis (premise) is correct. We then test the hypothesis by performing the experiment to see whether or not the results are as predicted. This deductive testing takes the form of "*If . . . then*" logic:

- **Observation:** My flashlight doesn't work.
- **Question:** What's wrong with my flashlight?
- **Hypothesis:** The flashlight's batteries are dead.
- **Prediction:** *If* this hypothesis is correct,
- **Experiment:** and I replace the batteries with new ones,
- **Predicted result:** *then* the flashlight should work.

FIGURE 4.8 Applying the scientific method to a campground problem.

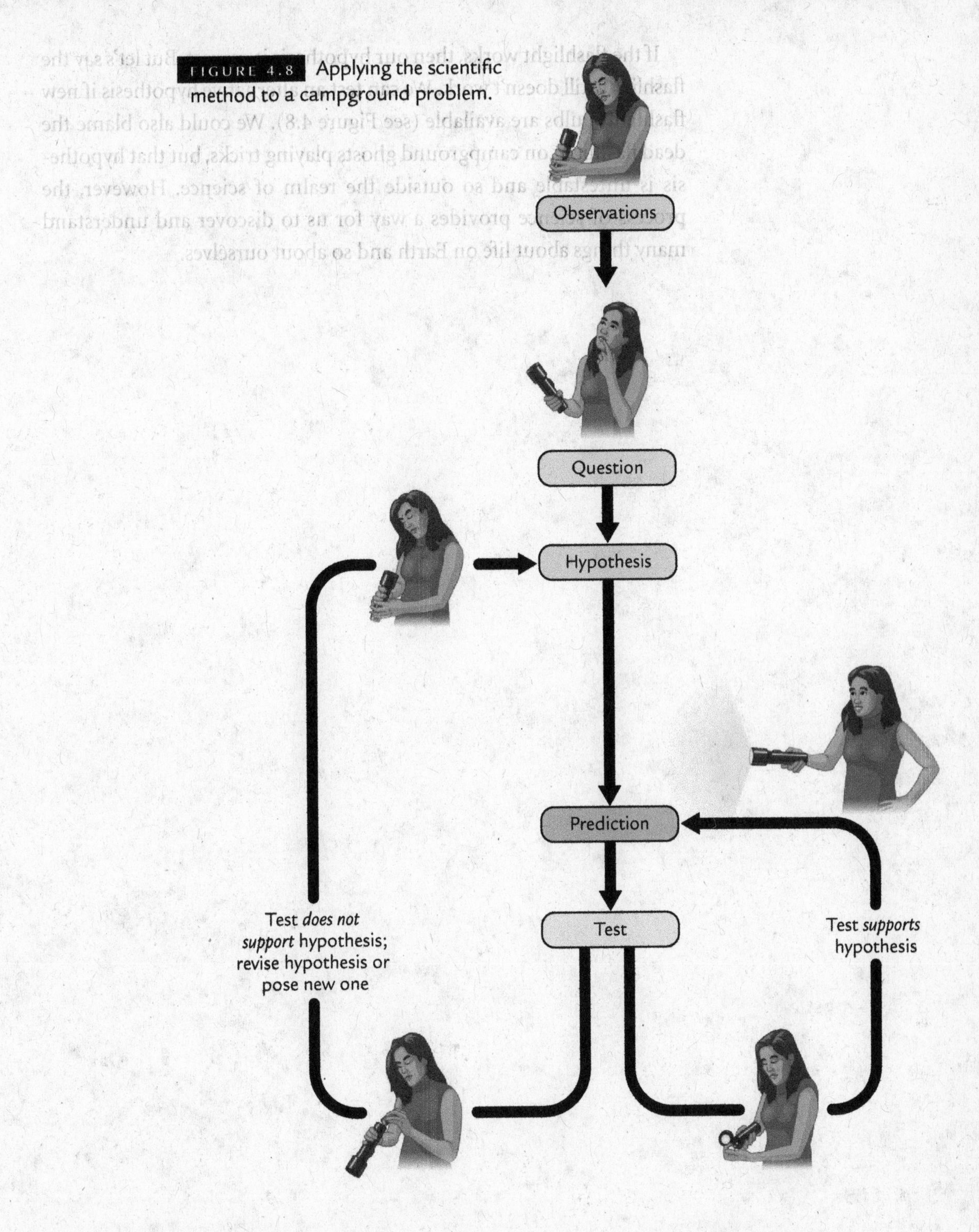

If the flashlight works, then our hypothesis is correct. But let's say the flashlight still doesn't work. We can test an alternative hypothesis if new flashlight bulbs are available (see Figure 4.8). We could also blame the dead flashlight on campground ghosts playing tricks, but that hypothesis is untestable and so outside the realm of science. However, the process of science provides a way for us to discover and understand many things about life on Earth and so about ourselves.

Final Stretch!

Now that you have finished reading this chapter, it is time to stretch your brain a bit and check how much you learned.

RUNNING WORDS

At the end of each chapter, be sure you have learned the language. Here are the terms introduced in this chapter with which you should be familiar. Write them in a notebook and define them in your own words, then go back through the chapter to check your meaning, correcting as needed. Also try to list examples when appropriate.

Chemical element
Atom
Molecule
Macromolecule
Organelle
Cell
Tissue
Organ
Organ system
Organism
Population
Community
Ecosystem
Biological system
Unicellular
Multicellular
DNA
Genes
Prokaryote
Eukaryote
Homeostasis
Negative feedback
Inheritance
Energy
Metabolism
Photosynthesis
Adaptation
Natural selection
Evolution
Biodiversity
Taxonomy
Domain Bacteria
Domain Archaea
Domain Eukarya
Theory
Scientific method
Hypothesis

WHAT DID YOU LEARN?

Try these exercises from memory first, then go back and check your answers, looking up any items that you want to review. Answers to these questions are at the end of the book.

PART A: ANSWER THE FOLLOWING QUESTIONS.

1. List, in order from most complex to simplest, the levels of the biological hierarchy of organization. ________

2. Examine your foot. It transmits your body weight to the ground and supports you in an upright position. How does your foot demonstrate that form fits function? ________

3. What is homeostasis and why is it important? ________

4. What substance transmits information about characteristics from parents to offspring? ________

5. Where do most living things on Earth ultimately get their energy from? ________

6. How are pesticide-resistant insects an example of natural selection in action? ________

7. Name the three domains and the four current kindoms of eukaryotes. ________

8. What is the difference between inductive and deductive reasoning? ________

PART B: MATCH THE TERMS ON THE LEFT WITH THEIR LEVEL OF BIOLOGICAL HIERARCHY ON THE RIGHT.

1. ________ nucleus	a) Macromolecule
2. ________ flock of birds	b) Organ
3. ________ seaweed	c) Organelle
4. ________ DNA	d) Organ system
5. ________ woodland	e) Ecosystem
6. ________ heart	f) Organism
7. ________ digestive system	g) Population
8. ________ nervous	h) Tissue

5

Chemistry

The Science of Stuff

When you complete this chapter, you should be able to:

- Understand the different states of matter.
- Describe atomic structure.
- Read and understand the Periodic Table of Elements.
- Explain ionic, covalent, and hydrogen bonding.
- Describe polar molecules and their unique characteristics.
- Discuss basic organic molecules.

Your Starting Point

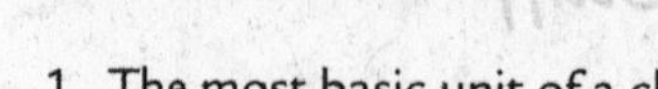

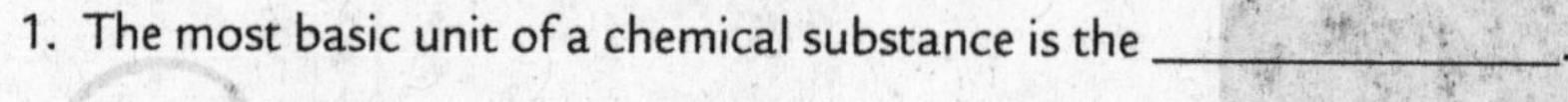

Answer the following questions to assess your chemistry knowledge.

1. The most basic unit of a chemical substance is the ____________.
2. Matter is defined as anything that ______________________________.
3. What are the three states of matter? ______________________________
4. What are the three most common subatomic particles? ______________________________
5. Which subatomic particles interact during chemical reactions? ______________________________
6. What is the molecular formula for water? ______________________________
7. What are three common types of chemical bonds? ______________________________
8. What happens in anabolic reactions? ______________________________
9. What is meant by *organic* molecule? ______________________________
10. Are proteins organic or inorganic? ______________________________

Answers: 1. atom. 2. has weight and occupies space. 3. Solid, liquid, gas. 4. Electrons, protons, neutrons. 5. Electrons. 6. H_2O. 7. Ionic, covalent, hydrogen. 8. Larger molecules are made. 9. Contains both C and H. 10. Organic.

Yes, we are going to tackle some basic chemistry, but relax—it is really not that difficult. Why do you have to learn chemistry? Biology is the study of all of the parts and materials in living organisms—all of the "stuff" that they're made of. And chemistry is the science that covers all of that stuff. In addition, all of the work done in cells involves chemical reactions. Chemistry is very much a part of our everyday lives. Some of you may have previously taken a chemistry class, but others will be new to this discipline. In this chapter, we'll explore basic chemistry concepts to give you a head start for your current class.

Recall from the last chapter our discussion of the biological hierarchy of organization. It is organized from the simplest level of organization to the most complex. The first three levels are part of chemistry so, as you see, chemistry forms the very foundation of biology (**Figure 5.1**).

Learning chemistry may seem tough at times because the terminology can be challenging. For example, just read the ingredient list on almost any food product. *Do we really eat all that stuff?* But don't let the words interfere with your understanding—much of chemistry is quite simple, even though it may not seem so at first. Here's an example. What do you know about a chemical compound called *dihydrogen oxide*? You probably know more than you realize—that is the technical name for something we usually call *water*!

What's the **Matter?**

Let's start with something you already know about: **matter**. All the "stuff" of which you are made is matter, and matter is defined as anything that

- has mass (or weight), and
- takes up space.

The terms **mass** and **weight** are often used interchangeably, but there is a difference. Mass refers to the actual physical amount of a substance. Weight takes into account the force of gravity acting on that mass. Consider astronauts. Each has a certain mass—the actual amount of material his or her body contains. Each astronaut is weightless during space travel when there is no gravity, but his or her individual mass does not

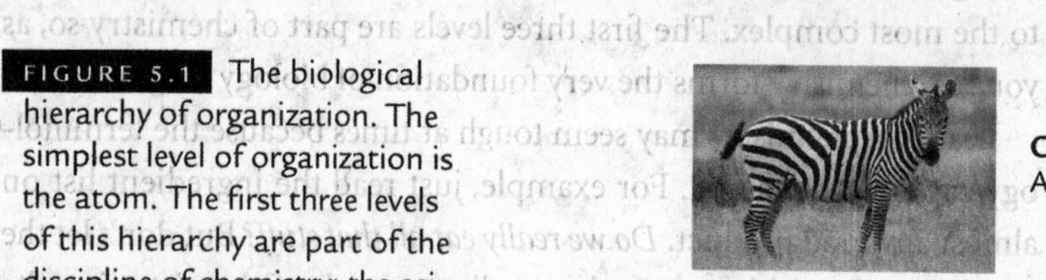

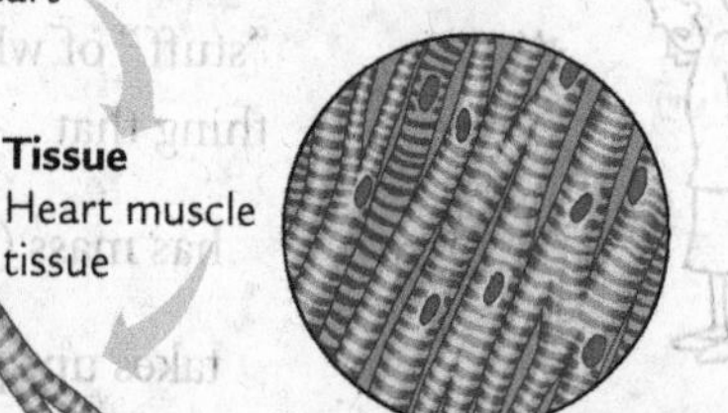

FIGURE 5.1 The biological hierarchy of organization. The simplest level of organization is the atom. The first three levels of this hierarchy are part of the discipline of chemistry: the science of matter. All of the other levels of organization are built upon this chemistry foundation.

change. For the sake of our discussion, matter can be defined using mass or weight, but mass is more precise. The second part of the definition of matter is that it takes up space. The space it occupies is called **volume**.

Matter typically exists in any of three physical states: solid, liquid, or gas. Once again let's consider water. What do we call the three states of water?

Solid: ____________ Liquid: ____________ Gas: ____________

Those should be easy. Solid water is ice, liquid water is water (duh!), and its gas form is vapor, or steam. See—you already know chemistry! Now, how can you change solid water to its gas form? ____________

____

When you add heat, which is a type of energy, ice melts to become a liquid. With enough heat, the liquid eventually boils to become vapor. If you collect the vapor and cool it, it will condense back to liquid. If you cool it enough, it will become ice (solid). As you can see, the three forms of matter are interchangeable.

✔ **QUICK CHECK**

What are the three states of matter? ____________

____

Answer: Solid, liquid, gas.

It's **Element**-ary, My Dear Watson!

All matter is composed of **elements**, which are the most basic chemical substances. Over 110 elements are recognized, and around 90 of these occur naturally on Earth. Some elements you likely know are iron, copper, silver, gold, aluminum, carbon, oxygen, nitrogen, and hydrogen. Some exist in pure form, such as helium and neon, but most occur combined with other elements.

For the most part, living organisms require only about 20 elements. By weight, 95% of the human body is composed of just four of these:

- carbon,
- hydrogen,
- oxygen, and
- nitrogen.

Each chemical is represented by a symbol, typically the first one or two letters of the element's name. If more than one element name begins with the same letter, the most common of these elements usually gets the single-letter symbol. The symbols for some of the elements, such as the four just listed, are quite logical. Others are less obvious. For example, the symbol for silver is *Ag*, but that is because it comes from the Latin word *argentums*, meaning *silver.* **Table 5.1** lists the names and symbols of some of the elements that are most important for life.

TABLE 5.1 Some of the important elements in living organisms.

Element name	Chemical symbol
Hydrogen	H
Carbon	C
Nitrogen	N
Oxygen	O
Sodium	Na
Magnesium	Mg
Phosphorus	P
Sulfur	S
Chlorine	Cl
Potassium	K
Calcium	Ca
Iron	Fe

TIME TO TRY

Several elements have names that begin with the letter C, so most of them use a two-letter chemical symbol. Try to match each of the following chemical names with their symbols. (*Hint: Recall that one of these is very common and is a major component of all living organisms, including the human body.*)

Your choices	Names	Symbols
________	Calcium	Cu
________	Chromium	C
________	Cobalt	Ca
________	Copper (*Latin = cuprum*)	Co
________	Carbon	Cr

Answers: Calcium = Ca; Chromium = Cr; Cobalt = Co; Copper = Cu; Carbon = C.

Chemical Carpentry: **Atomic Structure**

All chemical elements are composed of tiny particles called **atoms.** An atom is the smallest complete unit of an element—one atom of carbon, for example, is the smallest unit, or piece, of carbon that can exist. Two or more atoms can combine together to form larger structures called **molecules.** And simple molecules can join together to form more complex chemical structures called **macromolecules.** These include things like proteins, carbohydrates, DNA, and fats—many of the substances we associate with living organisms. But they all begin the same way—with atoms.

Atoms vary in size, weight, and how they interact with other atoms, but they all share some common characteristics. All are made of smaller units called **subatomic particles** that are arranged in a very precise manner. Although many subatomic particles are now recognized, the main ones of interest to us are **protons, neutrons,** and **electrons.**

THE NUCLEUS

The **nucleus** of an atom is not a structure. Instead, think of the nucleus as the area in the middle of an atom where some of the subatomic particles hang out. This can be confusing, because the nucleus of an atom is often referred to as if it is a structure. You should merely think of it as the atom's central region.

An atom's nucleus is where we find two types of relatively large subatomic particles called **protons** and **neutrons.** Protons and neutrons have a similar size and about the same mass. Protons are positively charged particles and may be designated as $\mathbf{p^+}$. Neutrons carry no electrical charge and they are, as their name suggests, neutral. Neutrons may be designated by $\mathbf{n^0}$, indicating they lack any electrical charge, or simply by **n.** All of the protons and neutrons in an atom are located in the nucleus.

ELECTRONS

Orbiting around the nucleus are the other major subatomic particles—the **electrons**—that are in constant motion. Electrons are very small and have almost no weight. They also carry a negative charge, and are often designated as $\mathbf{e^-}$. Because the protons are inside the nucleus, the nucleus always has a positive charge. However, the number of negatively charged electrons orbiting the nucleus always equals the number of protons at the nucleus. Thus the negative charges of the electrons exactly balance the positive charges of the protons. That means that any atom is, overall, neutral.

The number of e^- = the number of p^+. The atom is electrically neutral. ■

Electrons are never in the nucleus. Let's start with a simple image to get our bearings. Think of the rings of the planet Saturn. The rings never touch the planet itself. These rings can represent the paths of the electrons around the nucleus. Unlike Saturn's rings, however, the electrons do not travel in a nice, even, straight line along a single plane. Rather, they buzz about quite rapidly in multiple paths called **orbitals** that have different shapes and different orientations. For this reason, it is more accurate to envision a cloud of electrons that constantly circle the entire nucleus. A picture of an electron cloud does not show actual electrons. Instead, it is more like a map that shows the probability of where the electrons are at any moment.

PICTURE THIS

You arrive home late one evening, well after the sun has set. Your front light is on so you can see to put your key in the lock. You glance overhead and see a large cloud of insects swarming around the light. This is the basic image you should have of the electrons orbiting the nucleus of an atom (**Figure 5.2**). They are constantly in motion around the nucleus, but their individual paths vary.

Now that your skin is crawling from thinking about insects swarming overhead, let's get more specific. The movement of electrons around the nucleus is not as random as the movement of the insects around the light. It is actually a bit complicated, but for our purposes a simplified version will do. Electrons circle around the nucleus at different energy levels, each of which is called a **shell**, and these each have their own

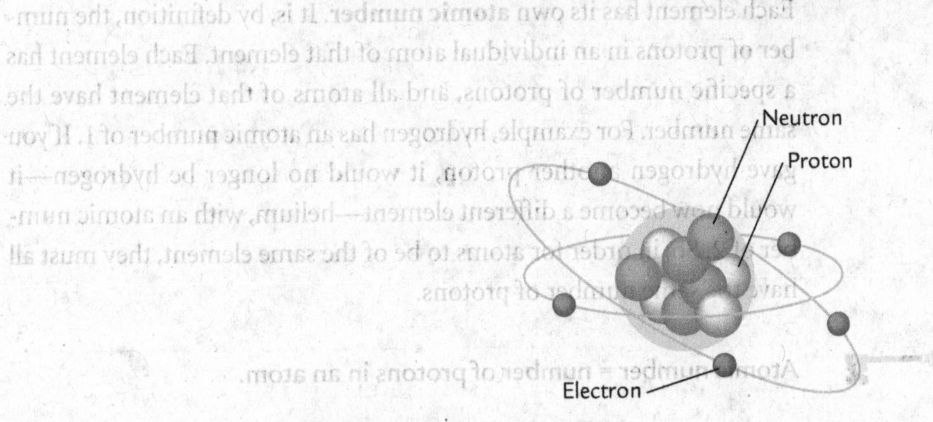

FIGURE 5.2 Electrons orbiting around the nucleus of an atom of boron.

paths around the nucleus. Each shell has its own range of distance from the nucleus and its own path shape, so you can think of each shell as being a specific part of the electron cloud. The first shell is closest to the nucleus. Electrons try to stay as close to the nucleus as they can, but the first shell can only accommodate two electrons. If an atom has more than two electrons, it must have more than one shell. Each additional shell is located a bit further from the nucleus. Electrons always fill the shells from the nucleus outward.

✔ **QUICK CHECK**

What are the three subatomic particles, and where is each located in an atom? ______________________________

____

____

Answer: Protons and neutrons are always in the nucleus; electrons are always orbiting around the nucleus.

ATOMIC NUMBER

Each element has its own **atomic number**. It is, by definition, the number of protons in an individual atom of that element. Each element has a specific number of protons, and all atoms of that element have the same number. For example, hydrogen has an atomic number of 1. If you gave hydrogen another proton, it would no longer be hydrogen—it would now become a different element—helium, with an atomic number of 2. So, in order for atoms to be of the same element, they must all have the same number of protons.

Atomic number = number of protons in an atom. ■

Now, recall that an atom is electrically neutral overall. This means that the number of positive charges from protons must be counterbalanced with an equal number of negative charges from electrons. In other words, the number of protons in an atom always equals the number of electrons. Because of this, if you know an atom's atomic number, you know not only how many protons it has, but also how many electrons it has—they are the same numbers! Let's try this.

TIME TO TRY

Nitrogen's atomic number is 7.

1. How many protons does an atom of nitrogen have? ___________
2. How many electrons does an atom of nitrogen have? ___________

If nitrogen's atomic number is 7, that tells you it has 7 protons. You know the number of protons must equal the number of electrons, so an atom of nitrogen will also have 7 electrons. The seven positive charges from the protons are balanced by the seven negative charges from the electrons, so the atom is electrically neutral.

The number of protons = the number of electrons in an atom. ■

ATOMIC WEIGHT OR MASS

An element's **atomic mass** refers to the total mass of a single atom of that element. But recall our earlier discussion about mass and weight: The terms **atomic mass** and **atomic weight** are used interchangeably. Because you are likely more familiar with weight than mass, we will stick with *atomic weight.* You need to recognize that

1. either of these terms can be used,
2. atomic mass is the more precise term, and
3. they have almost the same basic meaning.

For atomic weight we are really talking about how much a single atom weighs. Electrons are so tiny that they weigh almost nothing. Almost all of the atomic weight, then, comes from the combined weights of the larger protons and neutrons. But what do *they* weigh? Obviously atoms are too tiny to be weighed in pounds or ounces, so imagine trying to weigh a subatomic particle! Conveniently, scientists developed a unit of measurement called the **atomic mass unit (u)**, and 1 u is, to simplify it, about the mass or weight of one proton. Neutrons are almost the same size, so we assume that they have the same weight as a proton. This becomes amazingly simple! To determine an atom's weight, all you do is add its total number of protons and neutrons together. Because each of them weighs 1 u, the atomic weight is the total number times one, and you can simply ignore those tiny little electrons!

Atomic weight = number of protons + number of neutrons in an atom. ■

For all atoms of an element, the number of protons is constant. But the number of neutrons can vary, so atoms of the same element can have different atomic weights. For example, carbon (atomic number 6) has 6 protons, and it usually has 6 neutrons, so its atomic weight is usually 12. But some atoms of carbon have 7 neutrons, giving an atomic weight of 13, and some have 8 neutrons, giving an atomic weight of 14. Atoms of the same element that have different atomic weights are called **isotopes.**

WHY SHOULD I CARE?

Some isotopes are radioactive, meaning they emit certain types of energy. For this reason, some radioactive isotopes are used in medicine. The energy they emit can often be seen with special equipment. For example, the thyroid gland uses iodine to make certain hormones. If a patient might have a thyroid problem, a radioactive isotope of iodine (^{131}I) can be injected into the blood, then the clinician can use an imaging technique to monitor how well the thyroid is working. In another use, cobalt (^{60}Co) can be injected into an area where there is cancer, to irradiate the tumor cells.

Is That An Eye Chart or a **Periodic Table of Elements?**

Look at **Figure 5.3.** YIKES! It may look a bit scary at first, but that's only because you don't know how to read it. This is the **Periodic Table of Elements.** All chemical elements that are currently known are listed in this table, and more are added as they are discovered. The table is arranged in a specific manner that is quite useful, so we will take some time to explore it.

Look at **Figure 5.4a**. From the Periodic Table, we know that hydrogen's atomic number is 1, meaning it has 1 proton, which is shown in the center of the atom at the nucleus. This means it also has 1 electron, which is shown orbiting the proton, in the first shell. It rarely has any neutrons. Helium's atomic number is 2. The helium atom in **Figure 5.4b** has 2 protons and also 2 neutrons (so what is its atomic weight? ______). Helium also has 2 electrons, as shown. Next, look at lithium (**Figure 5.4c**), which has atomic number 3. You see it has 3 protons and

1A	2A	3B	4B	5B	6B	7B	8B	8B	8B	1B	2B	3A	4A	5A	6A	7A	8A
1 H 1.008																	2 He 4.003
3 Li 6.941	4 Be 9.012											5 B 10.81	6 C 12.01	7 N 14.01	8 O 16.00	9 F 19.00	10 Ne 20.18
11 Na 22.99	12 Mg 24.31											13 Al 26.98	14 Si 28.09	15 P 30.97	16 S 32.07	17 Cl 35.45	18 Ar 39.95
19 K 39.10	20 Ca 40.08	21 Sc 44.96	22 Ti 47.87	23 V 50.94	24 Cr 52.00	25 Mn 54.94	26 Fe 55.85	27 Co 58.93	28 Ni 58.69	29 Cu 63.55	30 Zn 65.41	31 Ga 69.72	32 Ge 72.64	33 As 74.92	34 Se 78.96	35 Br 79.90	36 Kr 83.80
37 Rb 85.47	38 Sr 87.62	39 Y 88.91	40 Zr 91.22	41 Nb 92.91	42 Mo 95.94	43 Tc (98)	44 Ru 101.1	45 Rh 102.9	46 Pd 106.4	47 Ag 107.9	48 Cd 112.4	49 In 114.8	50 Sn 118.7	51 Sb 121.8	52 Te 127.6	53 I 126.9	54 Xe 131.3
55 Cs 132.9	56 Ba 137.3	57 La 138.9	72 Hf 178.5	73 Ta 180.9	74 W 183.8	75 Re 186.2	76 Os 190.2	77 Ir 192.2	78 Pt 195.1	79 Au 197.0	80 Hg 200.6	81 Tl 204.4	82 Pb 207.2	83 Bi 209.0	84 Po (209)	85 At (210)	86 Rn (222)
87 Fr (223)	88 Ra (226)	89 Ac (227)	104 Rf (261)	105 Db (262)	106 Sg (266)	107 Bh (264)	108 Hs (269)	109 Mt (268)	110 Ds (271)	111 Rg (272)	112 — (285)	113 — (284)	114 — (289)	115 — (288)			

58 Ce 140.1	59 Pr 140.9	60 Nd 144.2	61 Pm (145)	62 Sm 150.4	63 Eu 152.0	64 Gd 157.3	65 Tb 158.9	66 Dy 162.5	67 Ho 164.9	68 Er 167.3	69 Tm 168.9	70 Yb 173.0	71 Lu 175.0
90 Th 232.0	91 Pa 231.0	92 U 238.0	93 Np (237)	94 Pu (244)	95 Am (243)	96 Cm (247)	97 Bk (247)	98 Cf (251)	99 Es 252	100 Fm 257	101 Md 258	102 No 259	103 Lr 260

FIGURE 5.3 The Periodic Table of Elements.

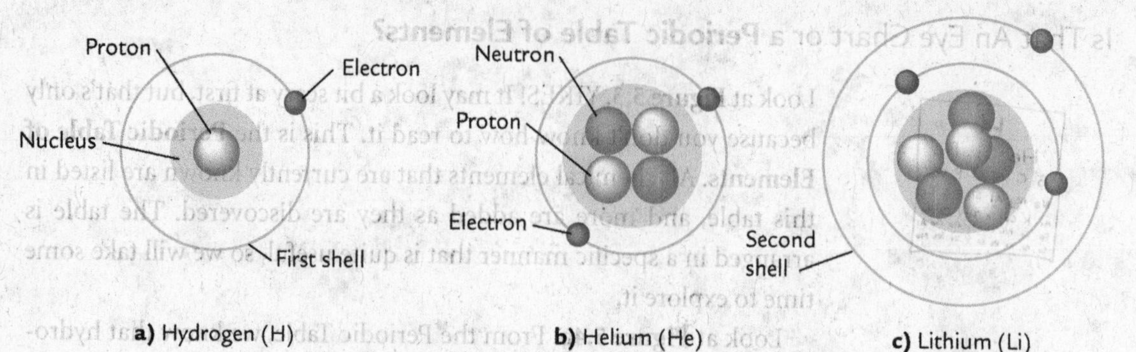

FIGURE 5.4 Illustrations of atoms of **a)** hydrogen, **b)** helium, and **c)** lithium showing the placement of the electrons around the nucleus. The first shell fills first and can hold only 2 electrons.

3 neutrons in its nucleus, and 3 electrons in orbit. The first shell can only hold 2 electrons, so a second shell is added to hold the third electron.

TIME TO TRY

In the illustrations below, assume that the gray sphere in the middle represents the nucleus and all neutrons and protons in it. Add the shells and electrons for each of the elements below. Boron is done as an example. Don't worry about the positions of the electrons; just draw the right number in the right shell.

Boron
Atomic number 5

Oxygen
Atomic number 8

Neon
Atomic number 10

Needs 5 e^- total.
2 e^- are in the inner shell,
the other 3 in the outer shell.

In your drawings, oxygen should have 2 electrons in its inner shell and 6 in its outer shell. Neon should have 2 electrons in its inner shell and 8 in its outer shell, giving it a full outer shell.

Now look at **Figure 5.5**. This is the square from the Periodic Table that represents carbon. You can see that each square of the Table tells you an element's chemical symbol, its atomic number, and its atomic weight. What do those three items tell you? ______________________

____

____

Remember, if you know an element's atomic number, you know how many protons are in its atoms. Once you know that, you also know how many electrons it has, because the number of protons and electrons is the same. But look at those atomic weights. Carbon's atomic weight is 12.01. Atomic weight equals the number of protons plus the number of neutrons. How can you have 0.01 (1/100th) of a proton or a neutron? You can't. Remember isotopes? The atomic weights shown in the Periodic Table are averages of samples that contain isotopes. You can round up or down to a whole number to figure out the typical number of neutrons.

How would you do that? If carbon has an atomic number of 6 and an atomic weight of 12.01, how many neutrons does it typically have? __________

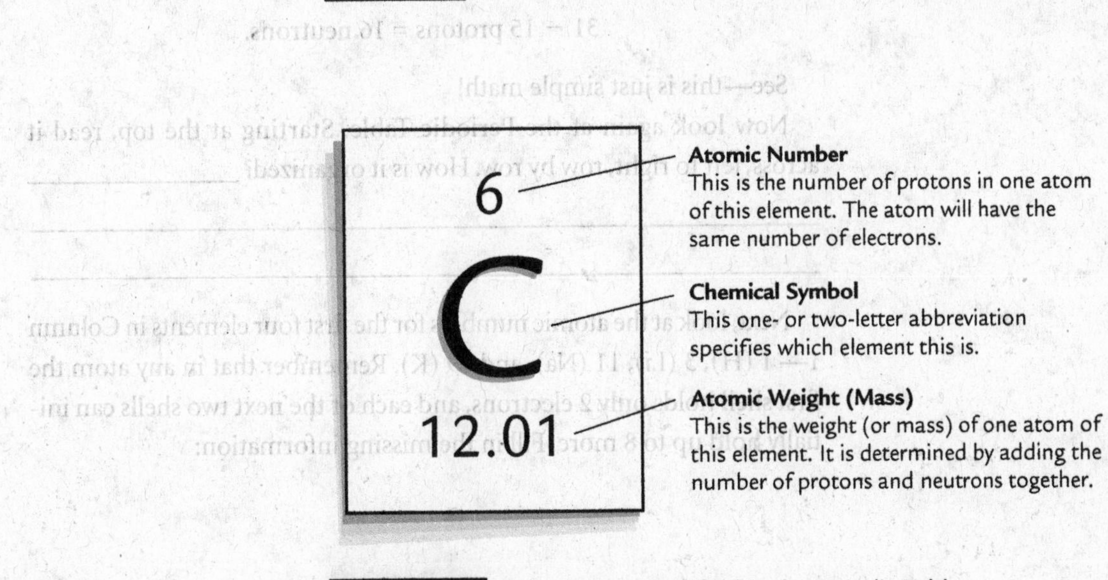
6
C
12.01
Atomic Number
This is the number of protons in one atom of this element. The atom will have the same number of electrons.
Chemical Symbol
This one- or two-letter abbreviation specifies which element this is.
Atomic Weight (Mass)
This is the weight (or mass) of one atom of this element. It is determined by adding the number of protons and neutrons together.

FIGURE 5.5 Information contained in the Periodic Table.

The correct answer is 6, because the atomic weight is closest to 12, of which 6 are protons. The rest are neutrons.

TIME TO TRY

Use the Periodic Table (Figure 5.3) to answer the following questions.

1. How many protons are there in an atom of calcium (Ca)? ______

2. How many electrons are there? ______

3. How many neutrons does a typical atom of calcium have? ______

4. How many neutrons does a typical atom of phosphorus (P) have? ______

You should see that calcium's atomic number is 20, so it has 20 protons and thus also 20 electrons. Its average atomic weight is 40.08, which rounds down to 40. Of that, 20 are protons, which leaves 20 neutrons. Phosphorus has an atomic number of 15, meaning it has 15 protons. Its average atomic weight is 30.97, closer to 31. So,

$$31 - 15 \text{ protons} = 16 \text{ neutrons.}$$

See—this is just simple math!

Now look again at the Periodic Table. Starting at the top, read it across, left to right, row by row. How is it organized? ______________________________

Next, look at the atomic numbers for the first four elements in Column 1—1 (H), 3 (Li), 11 (Na), and 19 (K). Remember that in any atom the first shell holds only 2 electrons, and each of the next two shells can initially hold up to 8 more. Fill in the missing information:

Element:	H	Li	Na	K
Electrons in its outer shell:	______	______	______	______

You should see that all elements in Column 1 have 1 lone electron in their outer shell. Hydrogen only has 1 electron. Lithium has 2 electrons in the first shell and 1 in the outer shell. Sodium has 2 in the first shell, 8 in the second, and 1 in the outer shell. Potassium has 2, 8, 8, and 1. If you do the same for the second column, you'll find that each element has 2 electrons in its outer shell. And you'll find that all the elements in the last column have 8 electrons in their outermost shells, except for helium, which only has 2 electrons.

The Periodic Table is organized by atomic number. The atomic number—the number of protons—increases from the left to the right, and from the top to the bottom. The Table is also organized into rows, called **periods**, and columns, called **groups**. Each row, or period, represents a shell of electrons. The first row has one shell, the second row has two shells, and so on. Each column, or group, represents how many electrons are in the outermost shell. This organizational approach is very simple to use for the first three rows of the Periodic Table, but becomes more complicated below that. Those complexities, however, are beyond the scope of our current discussion.

✔ **QUICK CHECK**

What information do you know from the period in which an element is found in the Periodic Table? ______________________

____

What information do you know from the group in which an element is found in the Periodic Table? ______________________

____

Answers: 1. The period tells you how many shells of electrons there are. 2. The group tells you how many electrons are in the outermost shell.

Bumper Cars and **Chemical Interactions**

Have you ever tried bumper cars? If not, you really should—it's a great way to release tension. Atoms interact with each other rather like bumper cars do (**Figure 5.6**). The first part of a bumper car that makes contact with another car is the outer rubber bumper. When two atoms come together, the first parts to make contact are always the electrons in the outer shells. The protons and neutrons are safely tucked away in the middle of the atom at the nuclei. So, the electrons in the outermost shell act as the "bumper" and determine how atoms interact with each other. We will discuss specifically how shortly.

Remember these two points:

1. The number of protons in all atoms of a particular element is constant.

2. The number of protons in an atom = the number of electrons in that atom.

From this, we see that all atoms of a particular element have the same number of electrons. Because the electrons determine their chemical activity, all atoms of a particular element will react the same way.

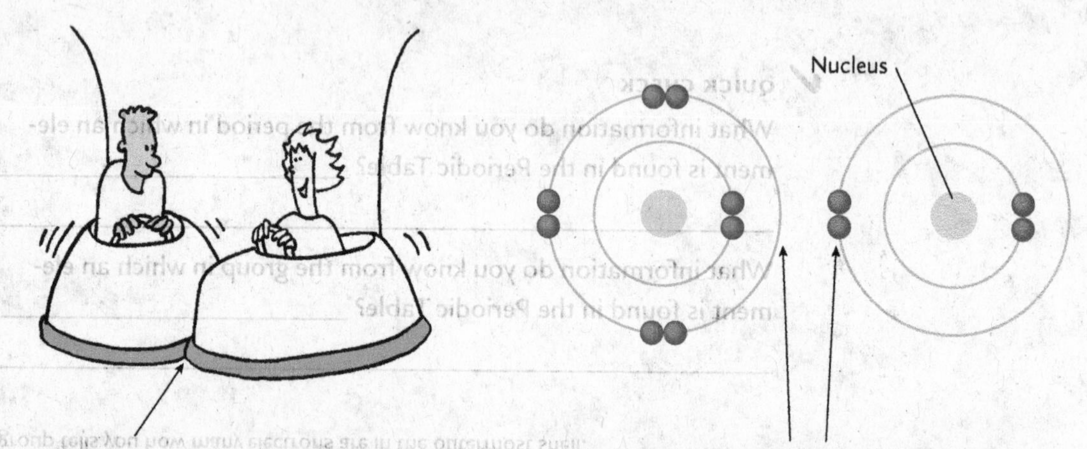
Nucleus
In bumper cars, the outer rubber bumpers of the cars make contact first. The riders inside the car never contact each other.
In atoms, the outer shell electrons make contact first. The protons and neutrons inside the nucleus never contact each other.

FIGURE 5.6 Chemicals interact somewhat like bumper cars. The electrons in the outer shells become the atoms' "bumpers" and determine the chemical's reactivity.

TIME TO TRY

Consider carbon in the Periodic Table.

What is its atomic number? ___________

How many protons does it have? ___________

How many electrons? ___________

Once you see that carbon's atomic number is 6, you know that it has 6 protons, and so it also has 6 electrons. Now draw the electrons for carbon around its nucleus.

You should see that carbon has 4 electrons in its second (outer) shell, and these are the electrons that will interact with other electrons.

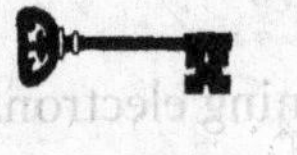

The electrons in an atom's outermost shell determine its chemical reactivity. ■

The Union: **Chemical Bonding**

As mentioned earlier, two or more atoms can join together through **chemical bonding** to form a molecule. Recall that the atom's electrons are arranged around the nucleus in one or more shells. These shells can have multiple subshells. The outermost of these subshells is called the **valency shell**. For simplicity, we will refer to this as the outermost shell. It can contain at most 2 electrons for helium, or 8 electrons for all other elements. If this outermost shell contains the maximum number of electrons, or is full, the atom is amazingly stable. It is said to be chemically

inert—it will not easily react with other atoms. It is "happy." All of the elements in the last column of the Periodic Table are inert.

On the other hand, atoms of elements in all of the other columns lack a full outermost shell. That means they are unstable and want to become stable. If it helps you remember this, think about life. When we are ful**filled**, we feel happy. If we are not happy, perhaps we feel something is missing from our lives, or maybe we feel we have lots of good to give and nobody to give it to. Now don't you feel sorry for those unfulfilled atoms?

An atom that does not have a full outermost shell of electrons is not stable, and it will react with other atoms to try to become stable. Unstable atoms can gain, lose, or share electrons with other unstable atoms until they become stable. That's how atoms interact. Let's explore this more deeply.

✔ **QUICK CHECK**

Under what circumstances is an atom stable? ______________________

____

____

Answer: An atom is stable when its outermost, or valency, shell is full, meaning it has 2 electrons for helium or 8 electrons for all other elements.

IONIC BONDING

When atoms become stable by losing and gaining electrons, electrons actually leave one atom's outermost shell and join the outermost shell of another atom. The atoms are now stable, meaning they each have a full outermost shell. However, gaining or losing electrons also changes the atoms in another way. Recall that atoms are normally electrically neutral—they have the same number of protons (+) and electrons (–). Once the electrons move, though, the atoms are no longer neutral because the protons and electrons are no longer balanced. An atom that gains an electron has one extra negative charge, and an atom that loses an electron is short one negative charge, making it positive.

All atoms that have gained or lost electrons carry an electrical charge and are called **ions**. These are designated with a $^{+}$ or $^{-}$ sign. For example, sodium tends to lose an electron and become a sodium ion, $\mathbf{Na^{+}}$. Chlorine tends to gain an electron, becoming a chloride ion, $\mathbf{Cl^{-}}$. Ions of opposite charges attract each other (*"Opposites attract"*). Whenever ions are formed, oppositely charged ions will join to form a molecule that is electrically neutral. When they join, they form a strong **ionic bond**—ions form ionic bonds.

Atoms that gain or lose electrons form ions, and ions of opposite charges form ionic bonds. ■

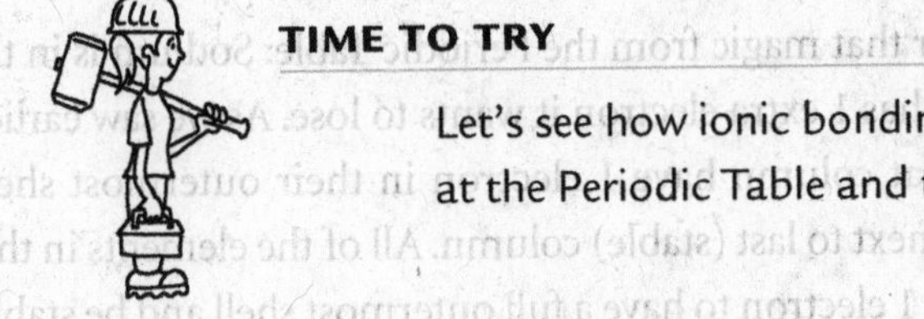

TIME TO TRY

Let's see how ionic bonding works, using sodium and chlorine. Look at the Periodic Table and fill in the following information:

	Sodium (Na)	**Chlorine (Cl)**
Atomic number:	____________	____________
Number of protons:	____________	____________
Number of electrons:	____________	____________

You know from the Periodic Table that neither of these elements is stable—they are not in the last column of the Table, so their outermost shells are not full. How can they become stable? ____________

____

Draw the electrons around the nuclei of each of these atoms (the shells are drawn for you):

Sodium

Chlorine

____

Sodium, with atomic number 11, should have 2 electrons in its inner shell, 8 in the second, and a single electron in its outermost shell. Chlorine, atomic number 17, should have 2 electrons in its inner shell, 8 in its second shell, and 7 in its outermost shell. That is pretty convenient—sodium has one too many and chlorine is short one. Sodium will lose its electron to chlorine, producing two ions, Na^{+} and Cl^{-}. Once the ions are formed, their opposite electrical charges will draw them together and they will form a strong ionic bond, creating a molecule of a substance called *sodium chloride*. You know it better as table salt!

Now remember that magic from the Periodic Table: Sodium is in the first column, so it has 1 extra electron it wants to lose. As we saw earlier, all elements in that column have 1 electron in their outermost shell. Chlorine is in the next to last (stable) column. All of the elements in that column need only 1 electron to have a full outermost shell and be stable.

What would you predict about calcium? ______________________

What about oxygen? ______________________

Calcium is in the second column, so it has 2 electrons in an outermost shell that wants 8. It is not stable. Oxygen is two columns short of being stable, so it needs 2 more electrons to fill its outermost shell and be stable.

For review, **Figure 5.7** shows how an ionic bond forms between lithium and fluorine. These elements are in the same columns as sodium and chlorine, so the process is the same except these new elements only have two shells of electrons. But, as you now know, in chemical interactions, only the electrons in the outermost shells are important.

✔ **QUICK CHECK**

How does an ionic bond form? ______________________

____

____

Answer: An ionic bond forms when atoms gain or lose electrons, forming oppositely charged ions that are then drawn together by their charges.

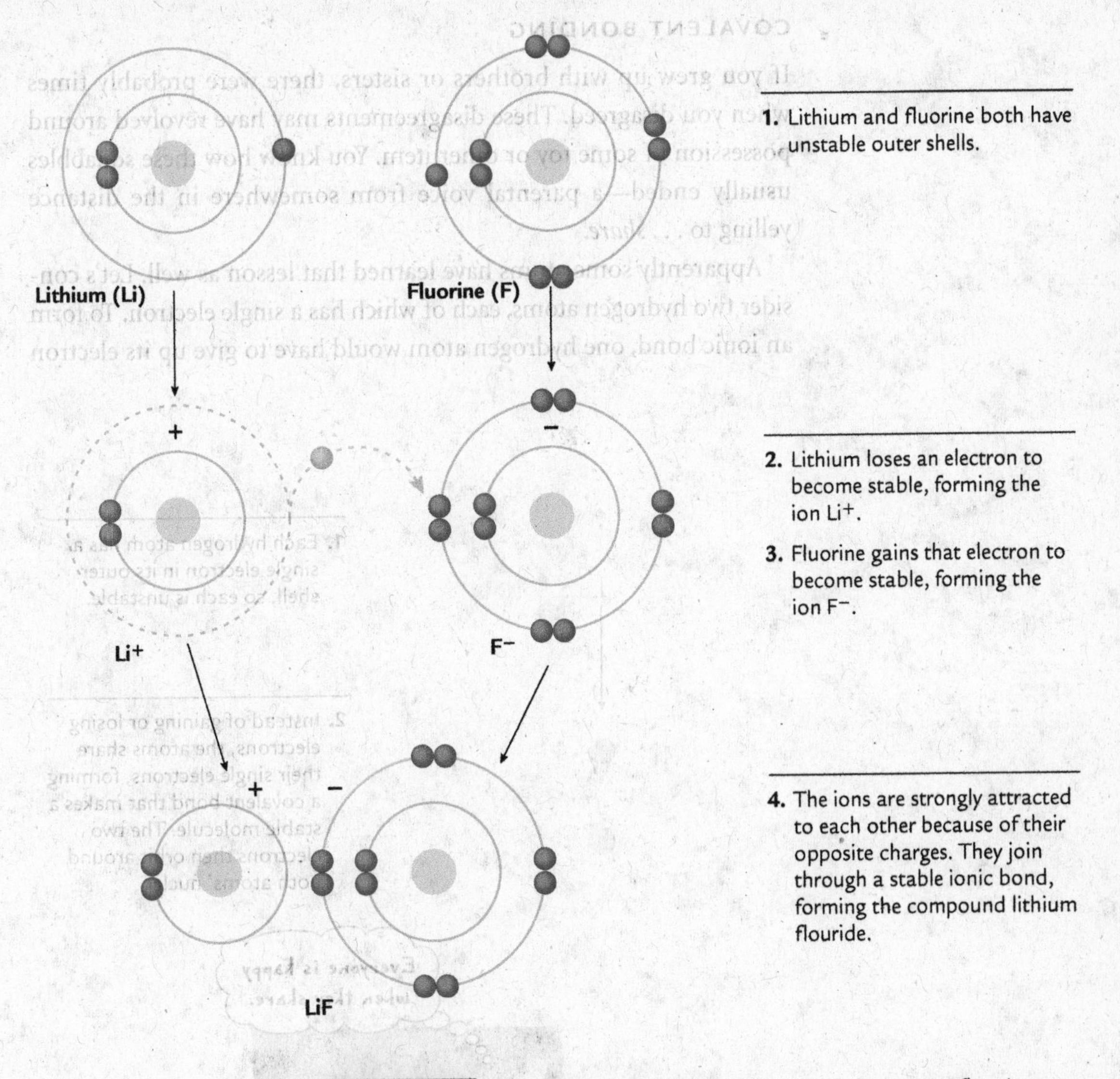
Lithium (Li)
Fluorine (F)
1. Lithium and fluorine both have unstable outer shells.
+
−
Li+
F−
2. Lithium loses an electron to become stable, forming the ion Li+.
3. Fluorine gains that electron to become stable, forming the ion F−.
+
−
LiF
4. The ions are strongly attracted to each other because of their opposite charges. They join through a stable ionic bond, forming the compound lithium flouride.

FIGURE 5.7 Ionic bonding. Lithium will lose an electron to fluorine, forming two oppositely charged ions. These ions are then attracted to each other and form an ionic bond, producing the compound called *lithium fluoride*.

COVALENT BONDING

If you grew up with brothers or sisters, there were probably times when you disagreed. These disagreements may have revolved around possession of some toy or other item. You know how these squabbles usually ended—a parental voice from somewhere in the distance yelling to . . . *share.*

Apparently some atoms have learned that lesson as well. Let's consider two hydrogen atoms, each of which has a single electron. To form an ionic bond, one hydrogen atom would have to give up its electron

1. Each hydrogen atom has a single electron in its outer shell, so each is unstable.

2. Instead of gaining or losing electrons, the atoms share their single electrons, forming a covalent bond that makes a stable molecule. The two electrons then orbit around both atoms' nuclei.

Everyone is happy when they share.

FIGURE 5.8 Covalent bonding. In covalent bonding, instead of engaging in a tug-of-war, two atoms share the electrons in their outer shells to become stable. Here a covalent bond between two hydrogen atoms is shown.

and another atom would have to gain it. But which will gain and which will lose? Neither will lose; instead, both hydrogen atoms can share their electrons. By combining them, the electrons will orbit around both nuclei together, and both atoms will be stable as long as they stay together. This type of bond is called a **covalent bond.** Recall that the outermost shell is called the valency shell and it contains the electrons that are interacting. The atoms that are sharing electrons in order to have full valency shells are said to be covalent (*co-* as in together or cooperating). **Figure 5.8** illustrates the formation of a covalent bond. In general, elements that are closer to the right or left side of the Periodic Table are more likely to form ionic bonds, and those closer to the middle of the Table are more likely to form covalent bonds.

✔ **QUICK CHECK**

How does a covalent bond form? ______________________________

____

Answer: A covalent bond forms when atoms share electrons to attain full outermost shells and become stable.

HYDROGEN BONDING

Although there are many types of chemical bonds, we will look at just one more. A **hydrogen bond** is a weak bond that can form between the hydrogen atoms in one molecule and some atoms in other molecules. Water is a classic example of this kind of bonding. Look at **Figure 5.9**. A water molecule has two hydrogen atoms and one oxygen atom. A hydrogen atom is tiny, with just 1 electron and 1 proton. An atom of oxygen is much bigger—it has 8 electrons, 8 protons, and 8 neutrons. Each hydrogen atom binds to oxygen by sharing its lone electron. Then the electron from hydrogen has to orbit around both nuclei—its own and also that of oxygen. The electrons from both hydrogen atoms will spend more time around the oxygen than around their own nuclei because of the size difference.

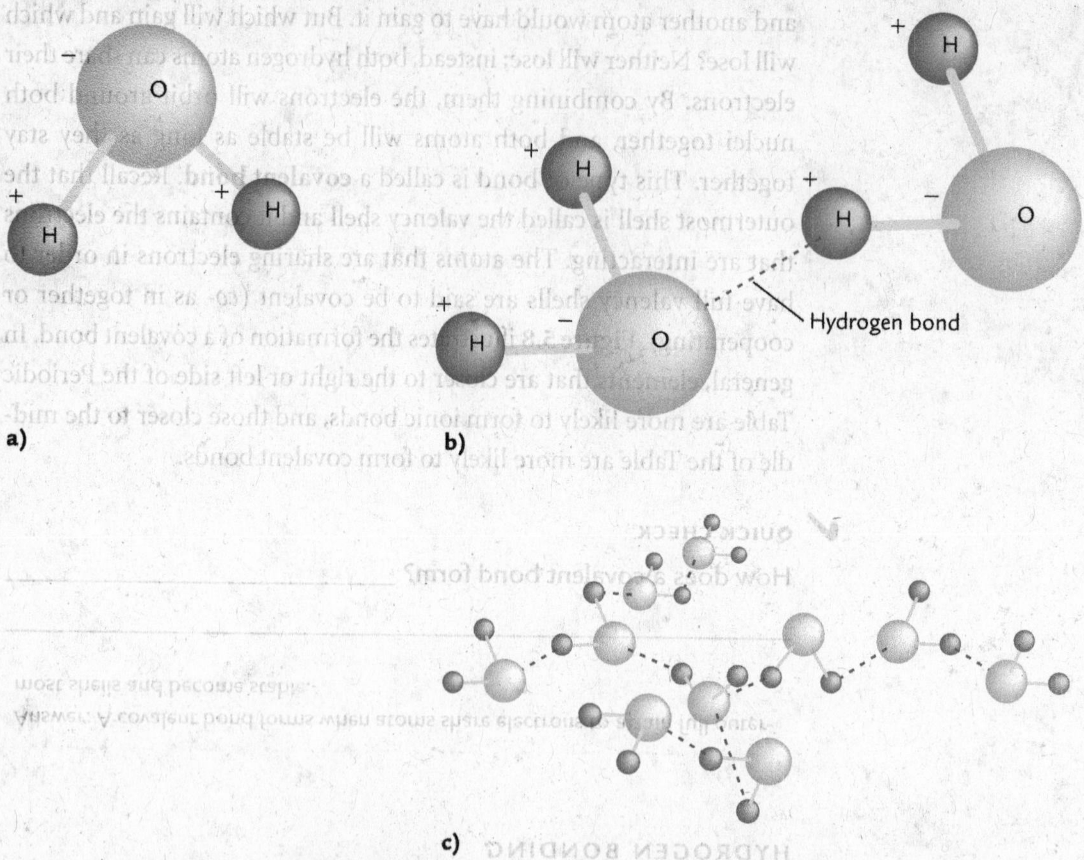

FIGURE 5.9 Hydrogen bonding. **a)** A water molecule is polarized—the hydrogen atoms tend to be slightly more positive than the oxygen. **b)** A hydrogen bond forms between a slightly positive hydrogen atom and a slightly negative oxygen atom. **c)** Hydrogen bonds between water molecules give water many unique properties.

PICTURE THIS

Prove this to yourself with Figure 5.9a. Using your finger to represent the path of the shared electron, move your finger at a steady pace to trace the path around one hydrogen atom and move from there to pass around the oxygen atom. Your finger is near the oxygen longer.

In addition, oxygen only shares 2 of its electrons, so it always has 6 around it, even without the shared electrons. So at any given time, there are more electrons around the oxygen than around the hydrogen. As a result, there is a slight imbalance in the electrical charge of the water molecule—the oxygen tends to be a bit more negative than the hydrogens because the electrons hang out there more. This slight charge imbalance is called **polarity.**

Polar molecules can form weak hydrogen bonds, as shown in **Figure 5.9b**. The slightly more positive hydrogen is attracted to the slightly more negative oxygen in an adjacent water molecule, so they form a weak bond. This is the hydrogen bond. It is kind of like striking up a conversation with a stranger in the checkout line at the grocery store. You happen to be standing close so perhaps you chat briefly, but you are not going to become lifelong buddies from that short, superficial exchange. Although they are weak, hydrogen bonds are important. They help shape many important molecules, such as proteins, your individual hairs (straight or curly), and your DNA.

Water has lots of hydrogen bonds, as shown in **Figure 5.9c**, giving it many unique characteristics. For example, ice floats because the hydrogen bonds cause the water molecules to spread out more in the solid form than they do in the liquid form. Water also has a high boiling point because of its hydrogen bonds.

TIME TO TRY

Let's examine the compatibility of polar and nonpolar substances.

1. Take any clear container, preferably one that can be closed. A plastic baggie will work. Fill it about a third full with water.

2. Add to that about half as much cooking oil. Try to get them to mix and observe what happens. ______________________________

   ____

3. Add a few drops of food coloring to the container and shake it well to mix.

▶

4. Is the food coloring polar or nonpolar (*Hint: Into which layer does it settle?*) ______________________________

5. Now find three harmless liquids and try mixing each of them with water. Avoid household cleaners that may be caustic or may react with each other if mixed.

Liquid tested	**Polar or nonpolar**
______________	______________
______________	______________
______________	______________

You likely found that your liquids, unless they were other oils, were polar, because they mixed with the polar water layer. Water is often used to make **solutions** because so many substances will dissolve in it. In a solution, one substance—the **solute**—dissolves in another substance—the **solvent**. Water is considered a universal solvent, and most chemical reactions in living organisms occur in water. Water readily dissolves polar molecules, but it causes nonpolar molecules to bunch together.

Water also has high *adhesion*, meaning it sticks to surfaces very well, and high *cohesion*, meaning its molecules stick to each other. Let's demonstrate that.

TIME TO TRY

You will need two pennies, alcohol, water, a dropper (or you can carefully use your finger), and a paper towel.

1. Place the pennies on a paper towel and examine them. Estimate how many drops of liquid you can put on a penny before it will spill over. ______________ drops

2. Start with the alcohol. Using the dropper or your finger, carefully place drops of the alcohol on the surface of the penny, counting each drop, until it overflows. How many drops of alcohol fit on the penny? ____________

3. Now use water and repeat this on the other penny. How many drops of water fit on the penny? ________

You should have been able to pile many more drops of water than alcohol on the penny because the water molecules stick to the penny and to each other much more than the alcohol molecules do. Because most beverages have a high water content, this property also allows you to fill a glass slightly higher than the rim (just don't try to put a lid on it or pick it up!)

✔ **QUICK CHECK**

Explain what is meant by *polar* molecule. ________________________________

____

Answer: A polar molecule is one in which there is an uneven charge distribution across the molecule, resulting in slightly positive and slightly negative charges.

MOLECULES AND COMPOUNDS

As mentioned earlier, when two or more atoms bind together, they form a molecule. If the atoms are from the same element, they form a molecule of that element; for example, O_2 is a molecule of oxygen. If the atoms are from different elements, the substance formed is called a **compound.** Water is a compound because it has two elements.

A molecule is described by a **molecular formula** that tells you what the molecule is made of. This formula includes the letter symbols for the elements and the number of atoms of each element that are present in the molecule. For example,

H_2O = water

O_2 = oxygen

CO_2 = carbon dioxide

CO = carbon monoxide

Note that the only difference between carbon dioxide and carbon monoxide is one atom of oxygen. We make carbon dioxide in our body and exhale it with every breath, whereas carbon monoxide is a deadly poison.

The molecular formula gives us limited information. It tells us how many pieces are in a molecule, but not how they are hooked together. For that, we can consult the **structural formula**, which is a simplified drawing of how the molecule is built. Lines in a structural formula represent chemical bonds (see Figure 5.9). Now look at **Figure 5.10**. This shows the structural formulae for three sugars: glucose, galactose, and

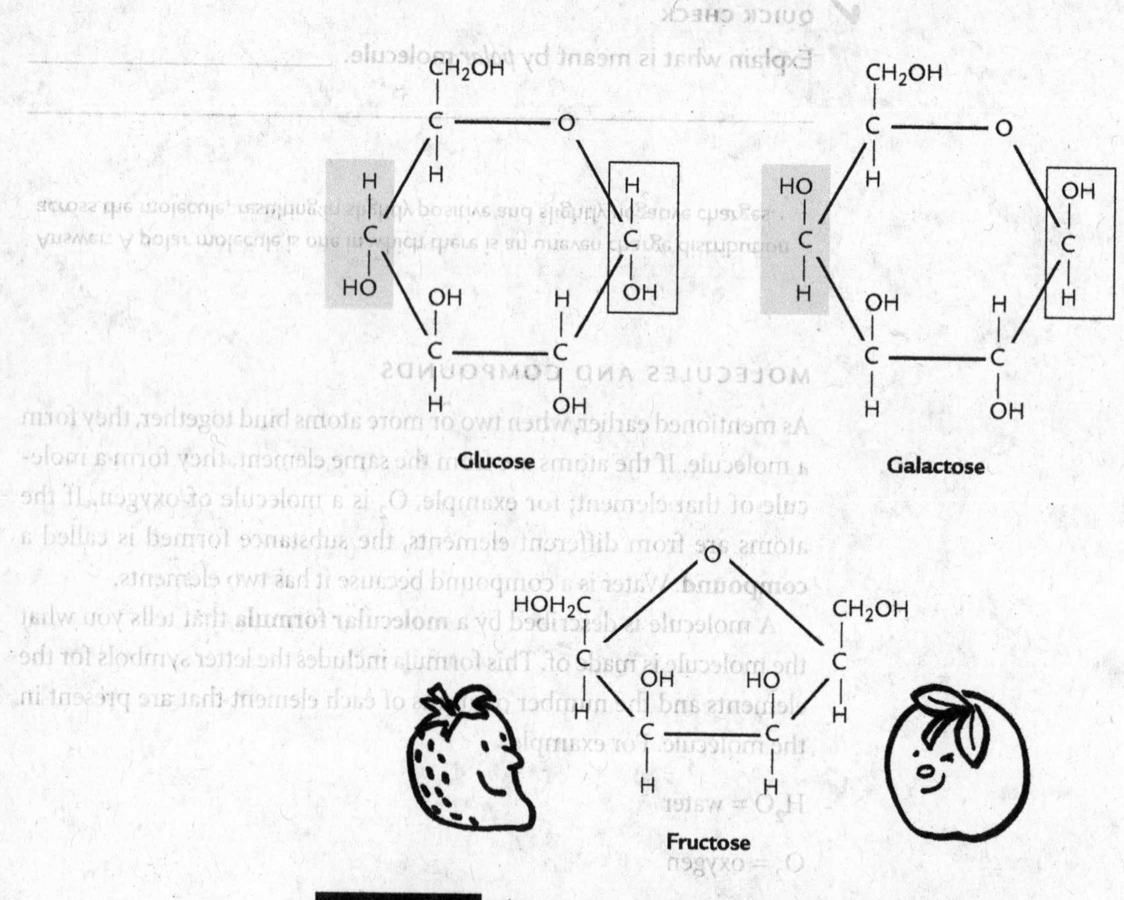

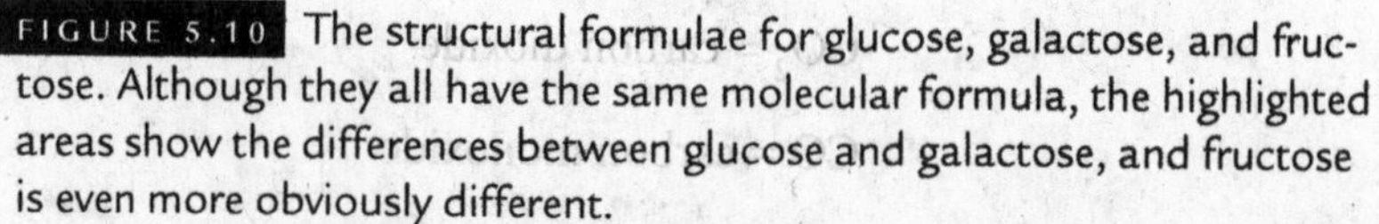
FIGURE 5.10 The structural formulae for glucose, galactose, and fructose. Although they all have the same molecular formula, the highlighted areas show the differences between glucose and galactose, and fructose is even more obviously different.

fructose. Glucose and galactose are quite similar, so the differences are highlighted. Fructose, also called fruit sugar, has an obviously different appearance.

TIME TO TRY

Look carefully at Figure 5.10 and fill in the following information.

	Glucose	Galactose	Fructose
Number of carbon atoms:	________	________	________
Number of hydrogen atoms:	________	________	________
Number of oxygen atoms:	________	________	________
Molecular formula:	C_H_O_	C_H_O_	C_H_O_

(*Hint: Use the numbers you wrote for each element above.*)

You can see that all three of these sugars have the same molecular formula: $C_6H_{12}O_6$. The structural formula provides more detailed information and is often more useful than the molecular formula, but the molecular formula is the most common method of describing molecules and compounds.

✔ QUICK CHECK

1. What is the basic difference between a molecule and a compound? ________________________________

____

2. What is the difference between a molecular formula and a structural formula? ________________________________

____

Answers: 1. A molecule is formed whenever two or more atoms join; if they are from different elements, the substance formed is a compound. 2. A molecular formula tells you how many atoms of each element are in a molecule; the structural formula shows how the atoms are connected.

Love Is just a **Chemical Reaction**

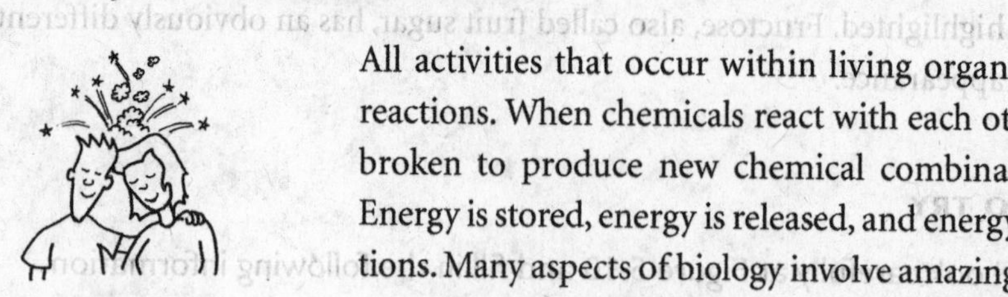

All activities that occur within living organisms start with chemical reactions. When chemicals react with each other, bonds are formed or broken to produce new chemical combinations or to release ions. Energy is stored, energy is released, and energy is used in chemical reactions. Many aspects of biology involve amazingly complex reactions that are meticulously controlled, whereas others are quite simple.

Chemical reactions are written in the form of a **chemical equation**, but instead of using an equals sign, we use an arrow. The substances to the left of the arrow are the **reactants**—the things that react together. The items to the right of the arrow are the **products**—the end result of the reaction. The arrow means "produces." For example, $Na^+ + Cl^- \rightarrow NaCl$ would be read "sodium ion plus chloride ion produces sodium chloride." Many reactions are **reversible**, meaning they can go in either direction. You can combine the ions we just mentioned to make salt, but salt can also break down to produce the ions. Reversible reactions are often indicated with a special double arrow symbol: $\rightleftharpoons$. Let's examine three basic types of reactions: synthesis, decomposition, and exchange.

Synthesis reactions are reactions that build. Another term for these is **anabolic reactions.** With that in mind, think about why athletes sometimes engage in illicit use of *anabolic* steroids—to *build* their muscles and strength. In synthesis reactions, two or more atoms or molecules combine to make a larger molecule. For example, proteins combine to build muscles. Small sugars combine to build large molecules of starch. This is done by forming new chemical bonds to hold the pieces together. Synthesis reactions are especially important in animals and plants for growth and repair processes.

Decomposition reactions are the opposite of synthesis reactions. During decomposition reactions, larger structures are broken down into smaller parts. These are also called **catabolic reactions**. For example, starch is broken down into smaller sugars. Salt is broken down to ions. These reactions are done by breaking chemical bonds. Decomposition reactions are especially important in animals for digesting food and producing energy.

Exchange reactions involve swapping pieces. Two or more molecules split apart and then recombine in a new way; for example, AB and CD separate, then recombine to form AC and BD. Exchange reactions allow an animal's body to receive and store chemicals in one form and then reuse them for multiple purposes.

WHY SHOULD I CARE?

All of the work done by the cells of living organisms—collectively referred to as *metabolism*—involves chemical reactions. We eat and breathe to bring in the necessary chemicals, then the body uses those molecules in an amazing array of chemical reactions that allow us to do virtually everything we do. All body processes rely on synthesis, decomposition, and exchange reactions. When you begin studying specific chemical processes, it may help to understand them if you think of what the outcome should be—are the reactions building, breaking down, or swapping? If you know the goal, you can more easily understand the reactions.

✔ **QUICK CHECK**

What are the three basic types of chemical reactions? ____________

____

Answer: The three basic chemical reactions are synthesis, decomposition, and exchange.

Is it **Organic?**

The term *organic* is used today to describe artwork, home décor, and how food is grown, among other things. In chemistry, however, it has a very precise meaning. By definition, **organic compounds** contain both carbon and hydrogen. They must have these two elements, and they usually have others as well. Most of your body is made of organic molecules.

That, Earthling, is what is meant by saying we are a carbon-based life form. The main categories of organic compounds are

- carbohydrates (sugars and starches),
- proteins,
- lipids (which include fats and steroids), and
- nucleic acids (DNA, which is your genetic material, and RNA, which assists DNA).

Any chemicals that do not contain *both* carbon and hydrogen are not organic, thus they are called **inorganic compounds**. These include carbon dioxide, oxygen, many acids and bases, salts, and water, which is our most important nutrient.

✔ **QUICK CHECK**

What is the difference between organic and inorganic molecules?

____

Answer: Organics contain both carbon and hydrogen; inorganics do not.

Our existence requires both organic and inorganic substances. These chemicals are constantly reacting, forming new bonds and breaking old ones, and they account for virtually everything you do and you are. Atoms and molecules are the very stuff of life, and the reactions between them are the processes of life. We hope now you can see that to achieve success in biology, chemistry truly does matter.

Final Stretch!

Now that you have finished reading this chapter, it is time to stretch your brain a bit and check how much you learned.

RUNNING WORDS

At the end of each chapter, be sure you have learned the language. Here are the terms introduced in this chapter with which you should be familiar. Write them in a notebook and define them in your own words, then go back through the chapter to check your meaning, correcting as needed. Also try to list examples when appropriate.

Matter
Mass
Weight
Volume
Element
Atom
Molecule
Macromolecule
Subatomic particle
Nucleus
Proton
Neutron
Electron
Orbital
Shell
Atomic number
Atomic weight/mass
Atomic mass unit (u)
Isotope
Periodic Table of Elements
Period
Group
Chemical bonding
Valency shell
Inert
Ion
Ionic bond
Covalent bond
Hydrogen bond
Polarity
Solution
Solute
Solvent
Compound
Molecular formula
Structural formula
Chemical equation
Reactant
Product
Reversible
Synthesis reaction
Anabolic reaction
Decomposition reaction
Catabolic reaction
Exchange reaction
Organic compound
Inorganic compound

WHAT DID YOU LEARN?

PART A: PROVIDE THE MISSING INFORMATION. CONSULT THE PERIODIC TABLE, FIGURE 5.3.

	Potassium	Iodine	Oxygen	Neon
Chemical symbol	______	______	______	______
Atomic number	______	______	______	______
Atomic weight	______	______	______	______
Number of protons	______	______	______	______
Number of electrons	______	______	______	______
Number of neutrons	______	______	______	______
Number of electrons in outermost shell	______	______	______	______

PART B: ANSWER THE FOLLOWING QUESTIONS.

1. Which subatomic particles are always in the nucleus? ______________________

____

2. Which subatomic particles are not included in the atomic weight? ______________________

____

3. Which subatomic particles have a positive charge? ______________________

____

4. What are the four main elements that make up most of the human body?

____

____

5. In the Periodic Table, what does the row (period) in which an element is positioned tell you about that element? ______________________

____

6. What does the element's column in the Periodic Table tell you? ______________________

____

7. Which elements are stable, and why?

____

____

8. How many electrons can the first shell hold? _____ The second shell? ______

9. A calcium ion has a $^{+}2$ electrical charge. How did this ion form to give it that charge? ________________________________

____

10. What is the basic difference between an ionic bond and a covalent bond? ________

____

____

11. Which gives you the most information: a molecular formula or a structural formula?

____

____

12. When you eat, the food is converted into small, simple molecules that can be absorbed into your blood. What type of chemical reactions does that involve?

____

13. If you get a paper cut, what type of chemical reactions will allow your skin to repair itself? ________________________________

____

14. What elements are always found in organic molecules? ________________

____

15. For each of the following, is the compound organic (O) or inorganic (I)?

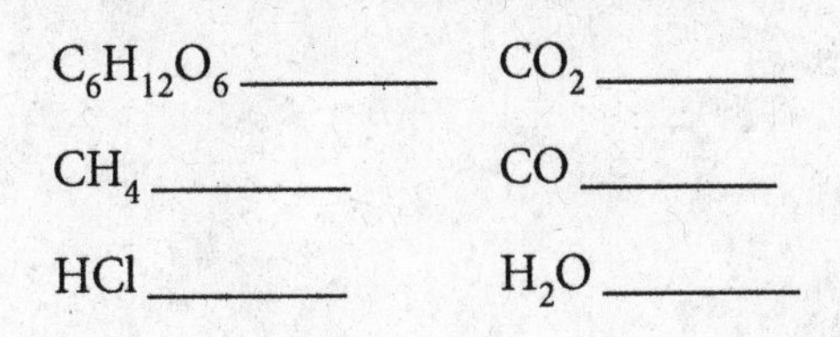

$C_6H_{12}O_6$ ________ CO_2 ________

CH_4 ________ CO ________

HCl ________ H_2O ________

6 Cell Biology

Life's Little Factories

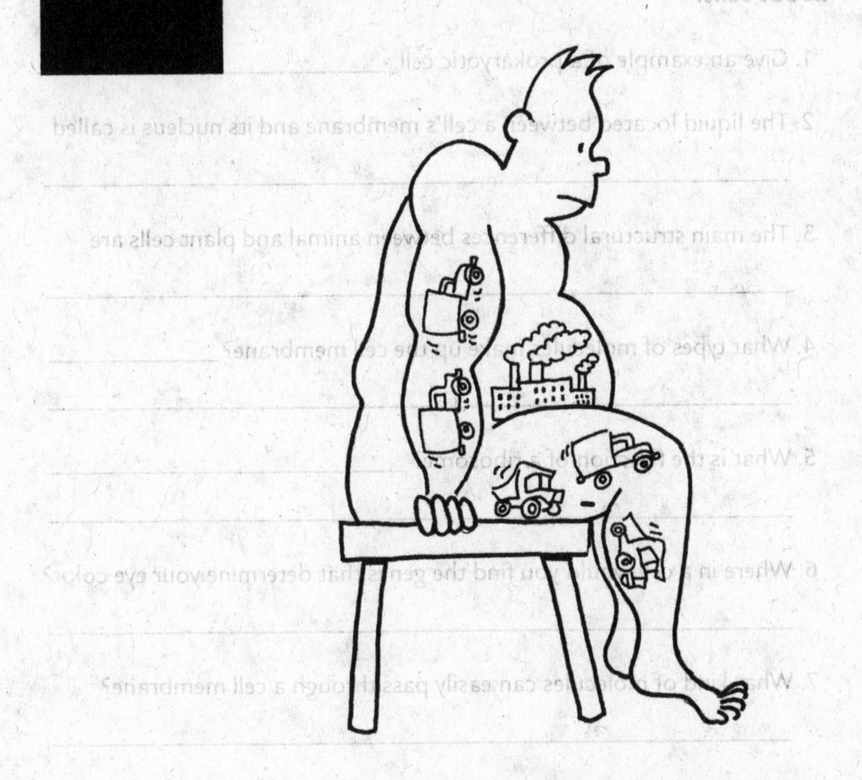

When you complete this chapter, you should be able to:

- Explain the cell theory.
- Distinguish between prokaryotic and eukaryotic cells.
- Describe the structures and functions of the cell membrane and cell organelles.
- Explain various movement processes that occur in cells.
- Describe the complete cell cycle and the basics of cell reproduction.

Your Starting Point

Answer the following questions to assess your knowledge about cells.

1. Give an example of a prokaryotic cell. ____________________

2. The liquid located between a cell's membrane and its nucleus is called ____________________.

3. The main structural differences between animal and plant cells are ____________________.

4. What types of molecules make up the cell membrane? ____________________

5. What is the function of a ribosome? ____________________

6. Where in a cell would you find the genes that determine your eye color? ____________________

7. What kind of molecules can easily pass through a cell membrane? ____________________

8. What is *diffusion*? ____________________

9. What moves during *osmosis*? ____________________

10. What happens during *mitosis*? ____________________

Answers: 1. Bacteria. 2. Cytoplasm. 3. Plant cells have cell walls and chloroplasts, whereas animal cells don't. 4. Phospholipids, proteins, cholesterol, and some carbohydrates. 5. Build proteins. 6. In the DNA in the nucleus. 7. Lipid-soluble molecules. 8. Movement of molecules from an area of higher concentration to an area of lower concentration. 9. Water. 10. The nuclear contents divide.

Good Things Come in Small Packages: **Cell Theory**

In the last chapter, we explored chemistry and the first few levels of the biological hierarchy of organization: atoms, molecules, and macromolecules. In this chapter, we continue our climb up the ladder (**Figure 6.1**) as we explore the next two levels: organelles and cells, which are where most of the chemistry occurs in our bodies. This chapter's focus is cell biology or cytology, which is the study of cells (*cyto-* = cell, *-ology* = study of).

In the 1600s, while examining slices of cork with a microscope, English naturalist Robert Hooke noticed that the cork was made of tiny chambers that reminded him of the cells in which monks lived in a monastery. From this observation, he coined the term **cell**, from the Latin word *cella*, meaning "storeroom" or "small container." Since then, vast amounts of cellular research have led to and continue to support a set of conclusions that are collectively referred to as the **cell theory**. Most sources acknowledge at least three major parts to this theory and some list as many as seven, so don't be surprised to see variation—all of these ideas are closely related. We shall consider the following five main principles:

1. All organisms are composed of one or more cells.
2. Cells are the basic structural and functional units of life.
3. All vital functions of an organism occur within cells.
4. All cells come from preexisting cells.
5. Cells contain hereditary information that regulates cell functions and is passed from generation to generation.

All organisms are composed of one or more cells. Whether the organism is a bacterium, fungus, plant, or animal, it is made of cells, and the cells of all these organisms show tremendous diversity. Cells come in many different shapes, sizes, and types. In fact, we humans have over 200 different types of cells in our bodies (see **Figure 6.2** on page 205). Red blood cells, or erythrocytes, are among our tiniest cells, measuring only about 2μm thick and 7μm in diameter. (Recall that a

FIGURE 6.1 Moving up the biological hierarchy of organization, we explore the organelle and cell, both covered in the discipline of cell biology.

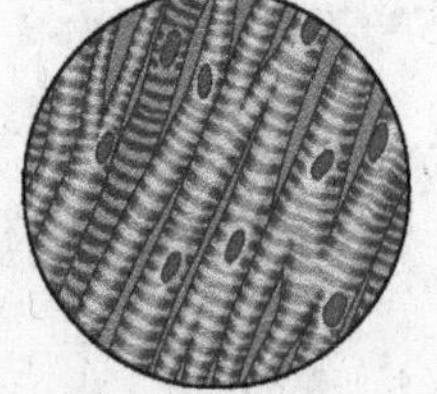

micrometer is only 1/1000 of a millimeter.) At the other end of the spectrum, a single neuron (nerve cell) may be up to a meter long! Sperm are minute compared to the ova that they fertilize. Yet, all of the diverse cells found in all organisms share striking similarities.

Cells are the basic structural and functional units of life. Cells are certainly the basic structural units—all organisms are built from them. But cells are also the functional units of life. Housed within its single cell, a **unicellular** organism has all of the structures and processes necessary to keep itself alive. Similarly, each cell in a **multicellular** organism is an independent living unit capable of maintaining itself. It takes in nutrients, uses them to make the molecules it needs to function, and harnesses energy for doing its work. While going about its daily business, the cell generates and also rids itself of waste. And, in most cases, it reproduces.

All vital functions of an organism occur within cells. Because each individual cell is alive and carries on all of its own life processes, it is no surprise that these same vital functions carried out at the higher levels of organization are still done by the cells. As you move up the ladder of complexity, life processes are performed within and around the cells. But as more cells are added, more organization is required to meet their

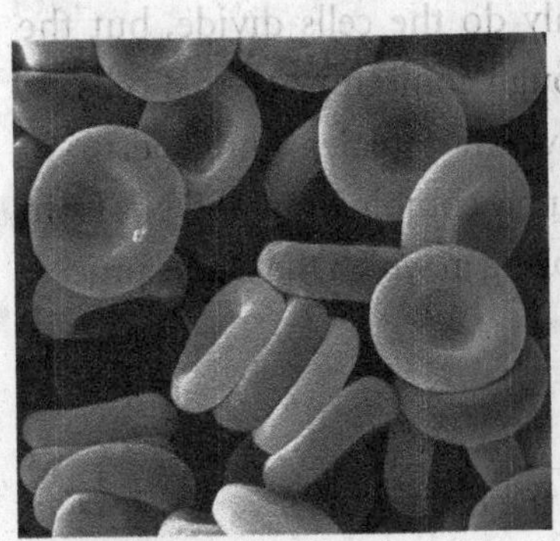

a) Red blood cells

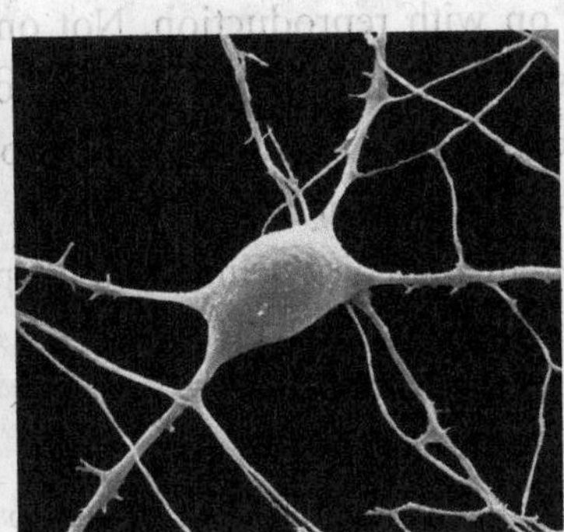

b) Nerve cell

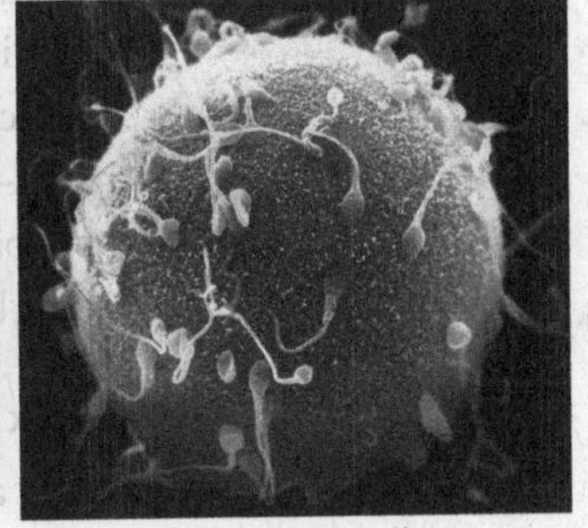

c) Egg cell and sperm cells

FIGURE 6.2 Human cells come in a variety of types and sizes. **a)** This scanning electron micrograph clearly shows that red blood cells are biconcave disks. **b)** A nerve cell (neuron) has numerous cell extensions that can be quite long, as shown in this scanning electron micrograph. **c)** A human ovum (egg cell) is covered by very tiny sperm cells at fertilization.

demands. As organisms increase in size and complexity, cells become more specialized and join together to share functions, forming tissues and eventually organs. For example, your digestive system brings in adequate nutrients for all of your cells, and your circulatory system ensures that all of your cells are well-serviced and able to communicate with each other. While your cells perform their highly specialized functions in complex organ systems, your nervous system carefully choreographs them. All the specializations in structure and organization keep your individual cells alive and functioning in a highly efficient and synchronized manner, while the cells go about their business of life, maintaining not only themselves, but also you.

All cells come from preexisting cells. Cells are capable of reproducing. The normal cell cycle, which we will discuss, ends with the cell dividing to produce two daughter cells. Each of the daughter cells is almost identical to the parent cell and will quickly take on its function. So the body parts of multicellular organisms are constantly being replenished as old cells die. Similarly, cells divide to replace other cells that may be lost during an injury.

Cells contain hereditary information that regulates cell functions and is passed from generation to generation. Inside every cell is the blueprint for life—the DNA that houses the cells' genes. These genes determine what work the cell will do, and these instructions for life are passed on with reproduction. Not only do the cells divide, but the organism itself can also reproduce through combining special cells. In multicellular organisms, gametes—ova in females and sperm in males—unite at fertilization, yielding a new combination of DNA from both parents, and producing a new offspring in which the whole process of life begins anew at many levels.

✔ **QUICK CHECK**

What are five principles that make up the cell theory?

____

____

____

____

____

A Cell by Any Other Name . . . **Prokaryotes and Eukaryotes**

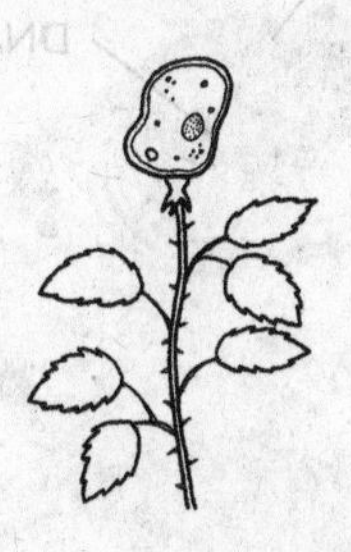

As mentioned earlier, cells can be quite diverse. Even within the human body we see tremendous variation in cell form, so imagine how much variety there is if we consider all living organisms! Despite this tremendous variation, all cells share at least three common characteristics:

1. They are enclosed in an outer **cell membrane**, which separates their internal environment from the external environment.
2. They are filled with **cytoplasm**, which is a mixture of substances in a liquid.
3. They contain deoxyribonucleic acid (**DNA**), which is the cell's genetic material.

As you already learned, all cells are either prokaryotic or eukaryotic. Let's look at their differences more closely (**Figure 6.3**). Prokaryotic cells don't have an enclosed nucleus. Instead, their DNA exists mostly in a loop and the area in which it tends to be located (but not enclosed) is referred to as the **nucleoid**. Additional DNA may be present in small loops, called **plasmids**, that can be transferred to other cells. Prokaryotes also lack most other internal cellular structures found in eukaryotes.

Answer: All organisms are composed of one or more cells. Cells are the basic structural and functional units of life. All vital functions of an organism occur within cells. All cells come from preexisting cells. Cells contain the hereditary information needed to regulate cell functions and it is passed on to the next generation of cells.

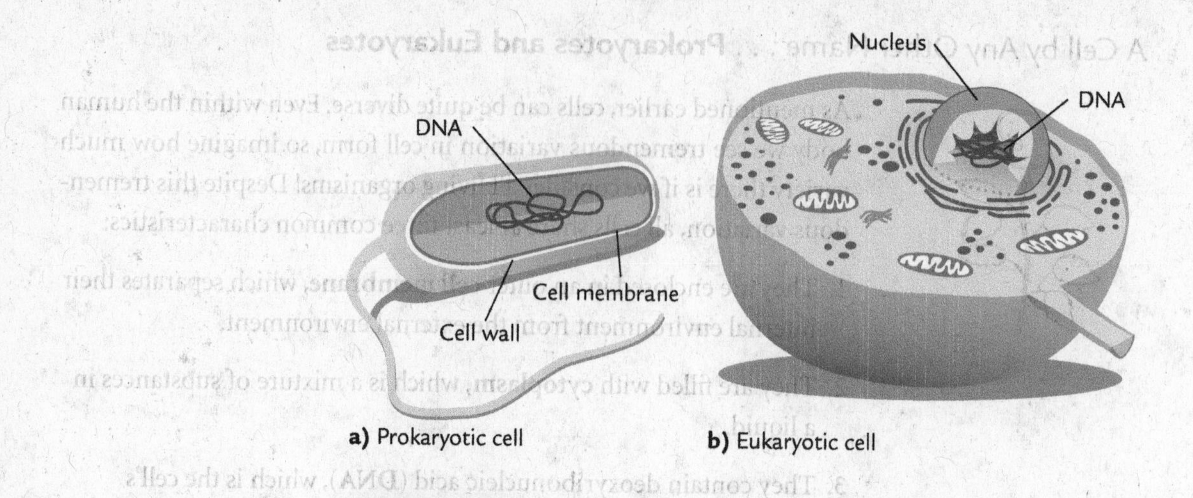

FIGURE 6.3 Cells are classified as either prokaryotic or eukaryotic.
a) Prokaryotes lack internal membranes and have no nucleus.
b) Eukaryotes have both.

WHY SHOULD I CARE?

Although animal and plant bodies are composed entirely of eukaryotic cells, prokaryotes are also important. For example, some bacteria enter our bodies through the foods we eat and take up residence in our colons (large intestines), where they live off our dietary leftovers. While consuming these materials, the bacteria actually produce vitamins that we use. They also, unfortunately, release methane gas in our guts that can cause us to be uncomfortable. Other bacteria can cause infection and illness if they enter our systems.

The cells of animals and plants, as well as many other forms of life, are all eukaryotic cells, or eukaryotes. Eukaryotic cells have a true nucleus and other organelles, as you learned earlier. Some of these organelles are built from membranes and others are not, but they all have specialized functions. All eukaryotes contain some of the same organelles, but specialized cells often have unique features as well.

All animal and plant cells are eukaryotic cells. ■

✔ QUICK CHECK

In which type of cell would you find a nucleoid, and how is it different from what would be present in the other type of cell?

Answer: A nucleoid is found only in prokaryotic cells and it is where the DNA loop is located. Eukaryotic cells, instead, have a membrane-bound nucleus that encloses their DNA.

Paper or Plastic? **The Cell Membrane**

All cells have certain requirements that must be met for them to give their peak performance. For example, they need the right amount of fluid, nutrients, water, and oxygen. Recall from Chapter 4 that **homeostasis** means maintaining a relatively constant internal environment (which you can think of as optimal working conditions). Doing so is not always easy.

One feature that helps achieve the goal of homeostasis is the cell membrane, also called the **plasma membrane**. It is simultaneously a container that holds a cell together and a physical partition that separates the cell's inside world from the outside, making it easier to maintain constancy inside the cell. The heading of this section, "Paper or Plastic," conjures up the image of a grocery sack. That image works for the concept of the cell membrane as a physical separation between the cell's inside and outside worlds. Clearly, this separation makes it easier to maintain homeostasis inside the cell even if conditions outside the cell vary.

But if the membrane was truly like an actual sack and formed a complete barrier, nutrients could not enter cells, nor could wastes and products manufactured by cells leave. Therefore, the cell membrane must have unique characteristics that allow some materials to pass through while blocking others.

The cell membrane is made primarily of **phospholipid** molecules. These molecules have phosphorous and other atoms at one end, forming what is called the **phosphate head.** This portion of the molecule is polar, or **hydrophilic.** Recall that term from the previous chapter's discussion of polar molecules. What does hydrophilic mean? ____________

____

Attached to one side of the phosphate head are two longer molecules called **fatty acid tails.** These tails are the main reason that the molecule is a lipid, and you should recall from the last chapter that lipids are nonpolar, or **hydrophobic.** What does that mean? ______________

____

To envision how a phospholipid molecule is organized, think about a brass brad—an old-fashioned paper fastener you likely used in school when you were young or that may even be built into some of your folders now (**Figure 6.4**).

TIME TO TRY

Let's try to figure out how the phospholipid molecules are arranged in a cell membrane.

1. What happens when a lipid comes into contact with water? Fill a clear container, such as a drinking glass or measuring cup, about 2/3 full of water. Watching closely, pour 1/2 teaspoon of cooking oil into the water. Describe what happens when the oil first enters the water, and where it ends. ______________

   ____

2. Get a slice of bread and cover half of its surface with a thick layer of peanut butter, which has a very high fat (lipid) content. Gently place one drop of water on the bare surface of the bread. What happens to the water? ______________

   ____

Gently place a drop of water on the peanut butter. What happens to the water? ______________________________

How do you think these observations relate to the organization of the phospholipid molecules in a cell membrane? ____________

____

Realize that a cell has an inside and an outside, and both contain water. In animals, the space outside of cells contains extracellular fluid that is mostly water. The cytoplasm inside cells also contains large amounts of water. The phosphate heads of the phospholipids interact fine with water, but the opposite ends of the molecules—the fatty acid tails—do not. As you saw when you placed oil in water, lipids in contact with water first form a ball, then settle into a single layer at the surface with one side in contact with the air, not water. In a cell, though, both surfaces are in contact with water. How can phospholipid molecules arrange themselves so the fatty acid tails do not contact water? ________

____

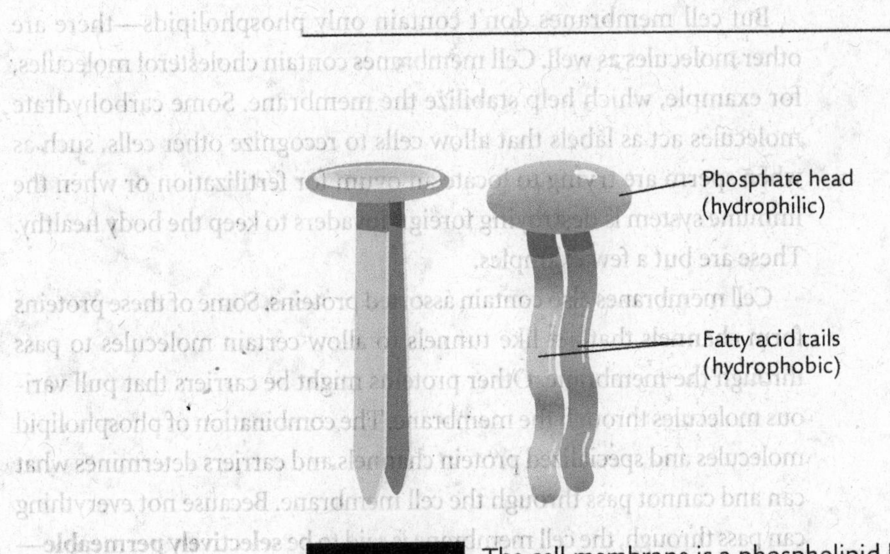

FIGURE 6.4 The cell membrane is a phospholipid bilayer. To understand the organization of a phospholipid molecule, think of a brass brad that is used to hold paper together. There is a head with two long parts hanging off of it. The head represents the phosphate "head" of our molecule, which is hydrophilic. The prongs that hang down represent the lipid "tails," which are hydrophobic.

Think about that partial peanut butter sandwich you made earlier. You should have seen the water readily enter the bare part of the bread, but ball up on the peanut butter. How could you organize bread and peanut butter in such a way that there are two major surfaces, or sides, and water will not be repelled from either one? ____________________

____

You guessed it—you have likely eaten it plenty of times. Consider the basic peanut butter sandwich. Assume you made it with two slices of bread, each smeared with a thick layer of peanut butter, then sandwiched together. You would have bread on both sides of the sandwich and peanut butter in the middle—two layers of it facing each other. This is the basic structure of the **phospholipid bilayer** (*bilayer* = two layers). The hydrophilic phosphate heads (bread) are arranged in two layers so that they face the water in the extracellular fluid and in the cytoplasm. The hydrophobic tails (peanut butter) are sandwiched in the middle, out of contact with the water. This is how the cell membrane is organized (**Figure 6.5**).

But cell membranes don't contain only phospholipids—there are other molecules as well. Cell membranes contain cholesterol molecules, for example, which help stabilize the membrane. Some carbohydrate molecules act as labels that allow cells to recognize other cells, such as when sperm are trying to locate an ovum for fertilization or when the immune system is destroying foreign invaders to keep the body healthy. These are but a few examples.

Cell membranes also contain assorted proteins. Some of these proteins form channels that act like tunnels to allow certain molecules to pass through the membrane. Other proteins might be carriers that pull various molecules through the membrane. The combination of phospholipid molecules and specialized protein channels and carriers determines what can and cannot pass through the cell membrane. Because not everything can pass through, the cell membrane is said to be **selectively permeable**—its chemical composition restricts the movement of some substances, so it is selective about what can pass through.

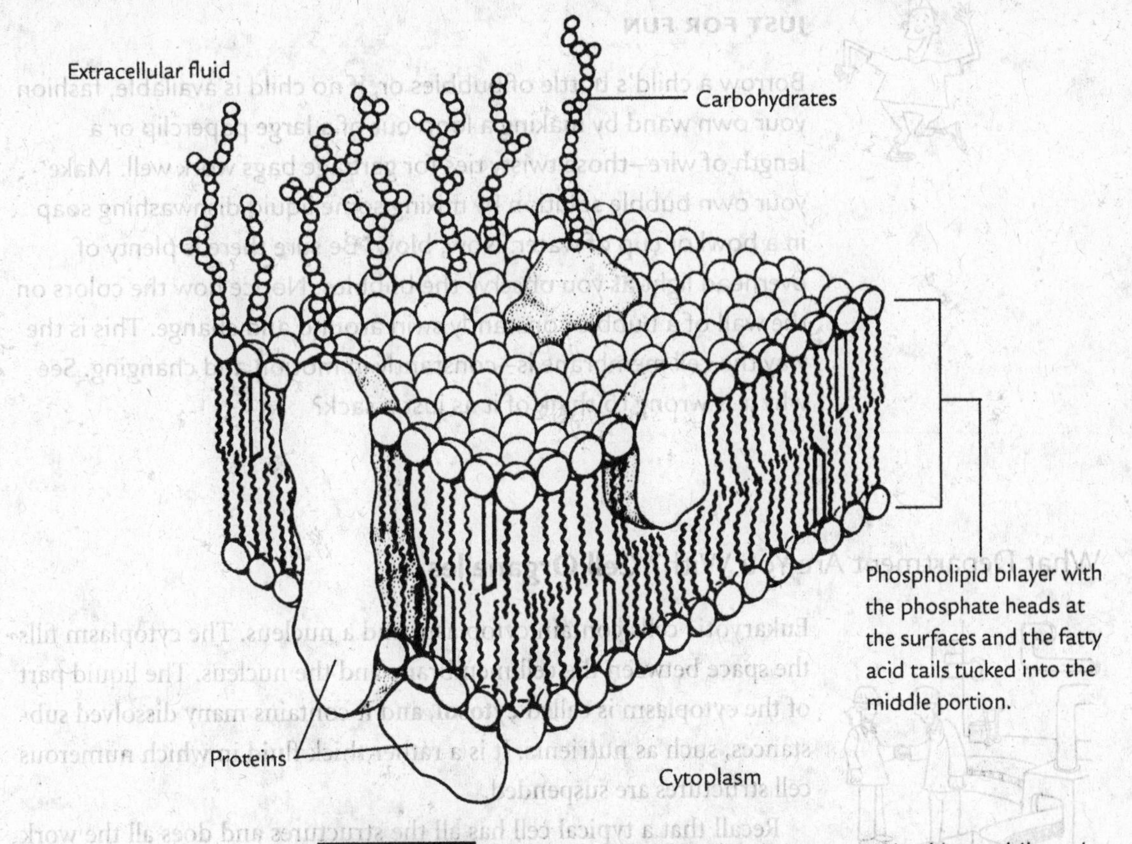

FIGURE 6.5 The phospholipid molecules are organized into a bilayer in which other molecules are embedded.

All of these assorted molecules are positioned throughout the cell membrane in what can be thought of as a sea of phospholipid molecules (see Figure 6.5). Imagine a child's pool filled with water, and floating in it are toy boats, inflatable toys, and a few children. The objects are free to move around in the water, and in fact they do. This is the nature of the cell membrane. This model, known as the **fluid mosaic model**, reveals a cell membrane that is very dynamic, moving, changing, and fluid in nature.

The cell membrane is composed of a phospholipid bilayer but also contains other molecules, and it is a very active structure. ■

JUST FOR FUN

Borrow a child's bottle of bubbles or, if no child is available, fashion your own wand by making a loop out of a large paperclip or a length of wire—those twisty ties for garbage bags work well. Make your own bubble solution by mixing some liquid dishwashing soap in a bowl or cup of water. Now, blow! Be sure there is plenty of overhead light as you observe the bubbles. Notice how the colors on the wall of a bubble constantly swirl around and change. This is the way the cell membrane is—constantly in motion and changing. See why it is wrong to think of it as just a sack?

What Department Are You With? **Cell Organelles**

Eukaryotic cells contain cytoplasm and a nucleus. The cytoplasm fills the space between the cell membrane and the nucleus. The liquid part of the cytoplasm is called **cytosol**, and it contains many dissolved substances, such as nutrients. It is a rather thick fluid in which numerous cell structures are suspended.

Recall that a typical cell has all the structures and does all the work needed to maintain life. Eukaryotic cells divide the labor among different specialized structures called organelles, each with a particular job. To understand this, think about a factory in which some product is manufactured. Inside this factory are many different departments, all involved in some way to make sure the factory is working properly and that the product is made and delivered. In a cell, the departments are the organelles. Although each has its own particular task, all organelles are coordinated and work together. The main structures of a eukaryotic cell are illustrated in **Figure 6.6.** Note that this figure shows that there are several main differences between plant and animal cells. For example, plant cells have rigid cell walls and green chloroplasts, neither of which is found in animal cells.

THE NUCLEUS

The **nucleus** is the largest organelle and it really has one specific task: It houses the DNA that contains genes. Each gene is essentially the instructions for how to make a specific protein. All cells in an animal's body

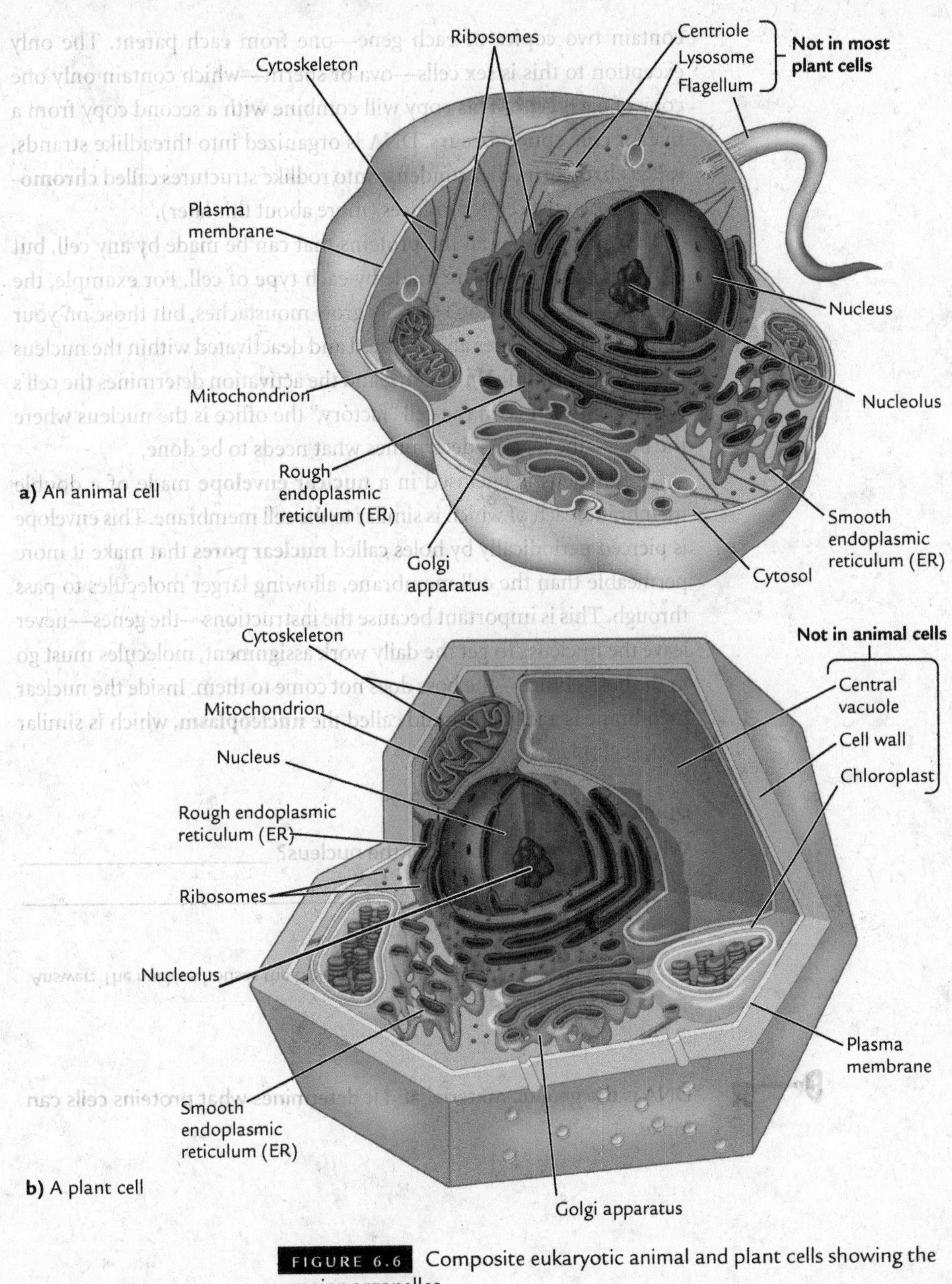

FIGURE 6.6 Composite eukaryotic animal and plant cells showing the major organelles.

contain two copies of each gene—one from each parent. The only exception to this is sex cells—ova or sperm—which contain only one copy of each gene. This copy will combine with a second copy from a mate if conception occurs. DNA is organized into threadlike strands, called **chromatin**, that condense into rodlike structures called **chromosomes** when the cell reproduces (more about this later).

DNA determines all the proteins that can be made by any cell, but only certain proteins are made by each type of cell. For example, the cells on your elbow don't usually grow moustaches, but those on your chin might! The genes are activated and deactivated within the nucleus based on many different factors, and the activation determines the cell's productivity. Thus, in the cell "factory," the office is the nucleus where the boss—the DNA—determines what needs to be done.

The nucleus is enclosed in a **nuclear envelope** made of a double membrane, each of which is similar to the cell membrane. This envelope is pierced periodically by holes called **nuclear pores** that make it more permeable than the cell membrane, allowing larger molecules to pass through. This is important because the instructions—the genes—never leave the nucleus. To get the daily work assignment, molecules must go to the boss's office—the boss does not come to them. Inside the nuclear membrane is a jellylike liquid called the **nucleoplasm**, which is similar to the cytoplasm.

✔ **QUICK CHECK**

What is the main function of the nucleus? ______________________

____

Answer: The nucleus houses the DNA.

DNA is the genetic material and it determines what proteins cells can make. ■

THE RIBOSOME AND PROTEIN SYNTHESIS

The DNA in the cell's nucleus has the instructions for all proteins that will be made in the cell, but the proteins are not manufactured in the boss's office. A separate cellular department—a tiny organelle called the **ribosome**—is responsible for assembling the protein during a process called **protein synthesis.** You can think of a ribosome as being the work bench on which the protein is built, rather similar to an assembly line. The workers who build the proteins there are called **RNA**, which stands for **ribonucleic acid.**

The instructions or work assignments from the boss—the DNA in the nucleus—are carried to the ribosome by a special worker, a molecule called **messenger RNA (mRNA)**. The mRNA carries the work assignment *to* the assembly line. Another worker, a type of RNA called **ribosomal RNA (rRNA)**, is always *at* the assembly line—it is part of the ribosome. This RNA is actually made inside the nucleus at a structure called the **nucleolus**, and it leaves the nucleus through those large nuclear pores.

Once the mRNA delivers the work order from the DNA to the ribosome, the specific proteins are made in the ribosome by linking together small molecules called amino acids. The amino acids are carried to the ribosomal assembly line by other workers called **transfer RNA (tRNA)**, then assembled according to the DNA's instructions. So RNA's job is to build the protein the DNA tells it to make. One type of RNA carries the instructions from the DNA to the ribosome, and the other two types of RNA are directly involved in building the protein according to DNA's specifications.

THE ENDOPLASMIC RETICULUM

Some ribosomes are free to float around the cytoplasm; others are attached to yet another organelle called the **endoplasmic reticulum**, or **ER.** The ER is an extensive network of membranous tubes and channels inside the cell. Under a microscope, ribosomes look like tiny dots, so when they are attached to the ER they give it a rough appearance; this is why it is called rough ER. The presence of ribosomes tells you that at least part of rough ER's job is protein synthesis. Smooth ER lacks ribosomes. Instead of making proteins, it is involved in making other materials (such as lipids), detoxifying potentially harmful substances, and transporting materials around the cell.

You can think of the ER as being like a system of workers who are always passing through various hallways, moving materials from one work station to another, and throughout the factory. Because the ER connects different parts of the cell, it also provides a communication network within the cell. Materials are brought into this system, moved around, changed, and turned into rough product here.

GOLGI APPARATUS

Under a microscope, the **Golgi apparatus** looks like a stack of flattened membranous sacs. This is the processing, packaging, and shipping department in cells. Products that have been made elsewhere in the cell, such as at the ribosomes or in the ER, are sent here to be finished. They are modified and put into their final forms. They are slapped with a molecular "shipping label"—a chemical tag that determines where they will go. They are packaged in some membrane that pinches off of the Golgi and surrounds the product, forming a saclike structure called a **vesicle.** Finally the products are shipped, some to other parts of the cell, some to the cell membrane, and some out to the great extracellular world beyond.

✔ QUICK CHECK

1. Where in a eukaryotic cell would you find ribosomes? ______

______

2. What is the functional relationship between the nucleus, nucleolus, ribosome, rough endopalsmic reticulum, and Golgi apparatus?

______

______

______

Answers: 1. Ribosomes may be free-floating in the cytoplasm or attached to endoplasmic reticulum. 2. The nucleus houses the DNA that has instructions for how to build proteins. Proteins are built at the ribosome, part of which is made in the nucleolus inside the nucleus. Many of the ribosomes are located on the rough ER. Proteins made at the ribosomes move into the Golgi apparatus to be processed, packaged, and shipped to their final destinations.

THE MITOCHONDRION

If the factory is to do its work 24/7, it needs a good power source. The **mitochondrion** is the cell's powerhouse. It provides a constant supply of energy to drive the work being done throughout the cell. Cellular energy in animals comes from food, where it is stored in the chemical bonds that hold the food's atoms together. Specialized chemical reactions in the mitochondria harness that energy and store it in a molecule called **ATP**—***a***denosine ***t***ri***p***hosphate. All cells, in any organism, use ATP directly for energy. Think of it as the electricity that powers cells. The more work a cell is doing, the more ATP it needs, and the more mitochondria it will have.

Mitochondria are unique organelles (**Figure 6.7**). They contain their own genetic information and can reproduce. These elongated organelles are enclosed by a double membrane, similar to the nuclear membrane. The outer membrane is smooth, but the inner membrane is highly folded. In general, such folding occurs to increase the surface area of the membrane, and increasing the surface area is the same as increasing the workspace. It is rather like expanding a department.

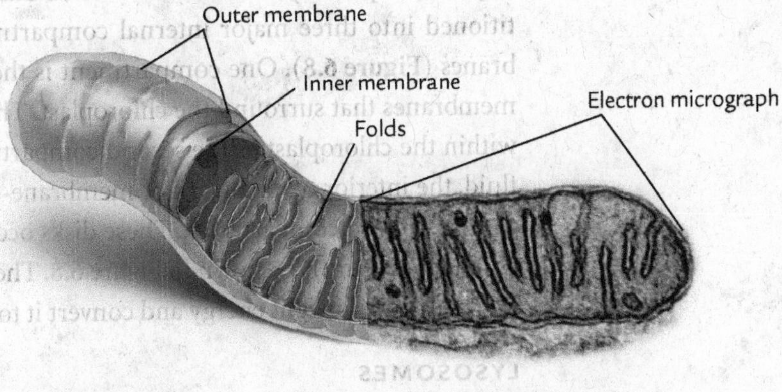

FIGURE 6.7 A mitochondrion. This illustration includes a drawing (left) blended into an actual electron micrograph.

TIME TO TRY

To understand how the foldings of the mitochondrial membrane increase the surface area, get a small plastic storage bag. This represents the outside membrane of the mitochondrion. Notice that it is smooth all the way around. If you put another smooth membrane inside of this bag, it would be about the same size. Now, take a larger plastic grocery bag and note how much larger it is. This will represent the inner folded membrane of a mitochondrion. Start folding it. Can you fold it enough to fit inside of the original bag? If you fold it enough, it should all fit inside. Now realize how much more membrane can be used if it is highly folded.

THE CHLOROPLAST

As you've learned, the energy that most living organisms exist on is produced by the process of photosynthesis, the conversion of light energy from the sun to the chemical energy of sugar and other organic molecules. Plants and other photosynthetic organisms are able to accomplish this because many of their cells, the ones that are green, contain unique organelles called chloroplasts. Like mitochondria, chloroplasts contain their own DNA and can reproduce.

Chloroplasts carry out the complex, multistep process that is required for photosynthesis. In order to do this, they are themselves partitioned into three major internal compartments by internal membranes (**Figure 6.8**). One compartment is the space between the two membranes that surround the chloroplast. The stroma, the thick fluid within the chloroplast, is the second compartment. Suspended in that fluid, the interior of a network of membrane-enclosed tubes and disks forms the third compartment. These disks occur in stacks called grana (singular granum), as shown in Figure 6.8. The grana are the structures that actually trap light energy and convert it to chemical energy.

LYSOSOMES

While the organelles are doing all this hard work in the cellular factories, they need some assistance in keeping their workplace tidy. **Lysosomes** are the cell's janitorial staff. A lysosome is a small membranous

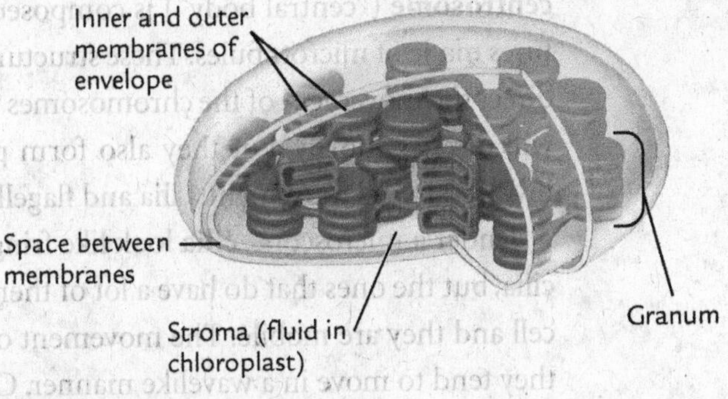

FIGURE 6.8 The chloroplast: site of photosynthesis.

bag containing strong digestive enzymes. Its main job is to break down materials. Some materials are brought into a cell and digested to provide basic building materials for the cell to use in its work. Other materials may be old, worn out cell parts or foreign material that invades cells. Lysosomes destroy these items so they neither harm nor clutter the cell's interior. A lysosome's job is to recycle what it can and get rid of the remaining garbage.

THE CYTOSKELETON

Remember that all of this work is going on inside a cell, in a liquid environment. The "building" needs a frame to hold it up. We discussed the outer partition—the cell membrane. But something is needed inside to hold the membrane out so the cell does not collapse on itself. The **cytoskeleton** is composed mostly of tiny tubes (**microtubules**) and filaments (**microfilaments**). These structures form a type of scaffolding that supports the cell and to which various organelles are attached. Although the name sounds like this structure is made of bone, the cytoskeleton is actually made of proteins. You can think of them as being the struts and beams that hold up the building, or cell.

CELL MOVEMENT: CENTRIOLES, CILIA, AND FLAGELLA

We have discussed how materials can move through a cell. But there are other movements associated with a cell, too. An area of the cell called the

centrosome ("central body") is composed of paired cylindrical structures made of microtubules. These structures are called **centrioles.** They direct the movement of the chromosomes when a cell reproduces, as we will discuss shortly, but they also form part of two other structures involved in cell movement: cilia and flagella.

Under a microscope, **cilia** look like fringe on a cell. Not all cells have cilia, but the ones that do have a lot of them. They are extensions of the cell and they are mobile. The movement of cilia is coordinated so that they tend to move in a wavelike manner. Cilia sweep materials over the outer surface of a cell, moving materials past the cell. For example, in the human respiratory tract the cilia help clear debris so it doesn't get into and clog the air sacs where oxygen enters our blood.

WHY SHOULD I CARE?

Cigarette smoke paralyzes the respiratory cilia for about an hour per cigarette. During that time, they cannot prevent the particulate material that we breathe in from reaching the lungs. Over time, more of this material clogs the small air sacs and begins to damage them. People who experience this must cough, especially upon arising in the morning, to try to clear the material that has accumulated in their lungs. This is the basis of *smoker's cough*.

PICTURE THIS

Imagine you are at the "big game" and the crowd is tossing a beach ball around. As this is going on, off in the distance you see the crowd start a "wave," where they stand and wave their arms overhead, then sit, group by group, all around the stadium. The wave approaches you just as the beach ball is heading your way. The wave passes you and so does the beach ball. It was carried away on the wave and you see it now making its way around the stadium, riding the wave. This is how cilia move materials across the surface of the cell—they beat in a coordinated manner, sweeping materials along.

In contrast to the cilia, a **flagellum** is a single, long, tail-like extension of the cell. In humans, these are found only on sperm cells. A flagellum whips back and forth to propel the sperm through the male reproductive tract and up into the female's tract in search of an ovum to fertilize.

We have reviewed the organelles of a typical eukaryotic cell. You will learn more detailed descriptions of these structures and their jobs as you proceed in your biology course. For now, you should have a general understanding that will give you a knowledge base from which to work. **Table 6.1** summarizes the organelles we have discussed and provides a quick review.

✔ **QUICK CHECK**

1. Flagella use tremendous amounts of energy to propel the cell forward. Which organelle would you expect to see in large numbers near the very busy flagellum? ________

2. How do the movements of cilia and flagella differ? ________

____

____

Answers: 1. Mitochondria. 2. Cilia are numerous and they beat in a wavelike manner to sweep materials across the cell surface; a flagellum is a single whiplike process that propels the entire cell.

Tunnels and Doorways: **Movement Processes**

We just discussed how cilia and flagella produce movement for cells, but these are not the only movement processes occurring. Atoms and molecules are always moving, both inside and outside the cell. Cell products packaged at the Golgi apparatus are shipped out of the cell to travel elsewhere in the body. Nutrients and building blocks are moved in from the outside. Waste products must leave. Let's examine how some of these movements occur.

BROWNIAN MOTION

Atoms and molecules constantly move in a random manner called **Brownian motion**. This is a rather nondirectional, jiggling movement.

TABLE 6.1 Summary of cell organelles and structures.

Organelle	Description	Function
Nucleus	Rounded larger membranous sac with pores.	Houses the DNA that directs cellular activities.
Chromatin	Relaxed strands of DNA in the nucleus.	Contain genes that determine what proteins can be made in the cell.
Ribosome	Small nonmembranous structure free in the cytoplasm or attached to ER.	Protein synthesis.
Nucleolus	Small body located in the nucleus.	Makes part of a ribosome.
Endoplasmic reticulum (ER)	Extensive network of membranous tubes and channels.	Protein synthesis (rough ER); lipid synthesis (smooth ER); detoxification; communication and transport system.
Golgi apparatus	Flattened stack of membranous sacs.	Processing, packaging, and shipping of cellular products.
Mitochondrion	Elongated membranous structure with highly folded internal membrane.	Powerhouse; harnesses energy from food molecules and stores it in ATP.
Chloroplast	Lens-shaped membranous structure with stacks of flattened disks called grana.	Photosynthetic organelle; converts energy of sunlight to chemical energy of sugars.
Lysosome	Small membranous sac.	Breakdown of unwanted materials; recycling of molecules.
Cytoskeleton	Meshwork of microtubules and microfilaments.	Provides support and structure to the cell interior; anchors organelles.
Centrioles	Cylindrical structures made of microtubules.	Direct movement of chromosomes during cell reproduction; form part of cilia and flagella.
Cilia	Small, numerous, hairlike processes that beat in a wave.	Sweep materials over the surface of the cell membrane.
Flagellum	A single whiplike tail.	Propels the cell forward.
Vesicle	Small membranous sac.	Contains materials entering or leaving a cell.

Remember those bumper cars we had fun with before? If you are in a car and someone hits you, their car bounces off yours in a new direction, while your car careens off in yet another direction. This is similar to how atoms and molecules move. Because they are constantly in motion, you can imagine how the particles will occasionally bump into each other and ricochet away.

CONCENTRATION GRADIENTS AND EQUILIBRIUM

Although Brownian motion is random, other movement processes are not. Before we discuss these movements, though, we need to understand the concepts of concentration gradients and equilibrium. You can think of concentration as referring to how crowded together molecules are—the more crowded they are, the more concentrated they are. A **concentration gradient** exists whenever there is a difference between the concentrations of the molecule in two areas (**Figure 6.9a**). Perhaps your vehicle has gradient glass in its windshield—it is tinted with more color at the top than at the bottom, which means there is a *gradient* in the color.

If there is about equal space between all of the molecules, we say they are at **equilibrium (Figure 6.9b)**. Think about standing in an elevator. If there are only two of you on the elevator, you likely stand at opposite sides. Add two more people and you likely stand one in each corner with

a) A concentration gradient

b) Equilibrium

FIGURE 6.9 **a)** A concentration gradient exists when there is a difference in the concentration, or spacing, around molecules in two different areas. **b)** Equilibrium exists when the molecules are spaced about evenly.

about equal distance between yourselves. As you add more and more people to the elevator, you become more closely crowded and are more likely to bump into each other. And if you do collide, you respond by moving away, always trying to maximize your personal space. That is how atoms and molecules move, but not by conscious decision. They move by basic laws of science. Let's explore how.

SIMPLE DIFFUSION

One of the most basic types of molecular movement is **simple diffusion**. This is how oxygen enters cells and carbon dioxide leaves them, for example, and it is critical for sustaining life. Simple diffusion occurs when molecules move from an area of higher concentration to an area of lower concentration. In other words, they move from where they are more crowded to where they have more room. You can think of it simply as the molecules spreading out. Whenever there is a concentration gradient, molecules will spontaneously move *down the concentration gradient.* That means they move from where they are most concentrated, or crowded, to where they are least crowded. You are already familiar with diffusion. Consider, for example, baking brownies. The molecules that produce the yummy smell quickly diffuse through the air in your house so that any visitors know you have a treat to share!

With individual molecules the motion is random, but like people in an elevator, the more molecules that are present in an area the more likely it is that they will bump into each other and bounce away. There will be more collisions in an area where the molecules are more crowded, sending them skidding off. They will ricochet less as they move into areas where there are fewer molecules to bounce off of. In time the molecules, like the people on the elevator, will have moved around enough to have almost equal distance between them. This is equilibrium. However, unlike the people on the elevator carefully maintaining their positions, molecules at equilibrium do not stop. They continue moving, but all molecules experience about the same number of collisions and they maintain a fairly even spacing.

The rate at which molecules diffuse varies under different conditions. Molecules diffuse faster when there is a greater concentration gradient between the two areas. Molecules in high concentrations will diffuse faster than those in lower concentrations. Smaller molecules also move faster than larger molecules, and temperature will alter the diffusion rate as well.

TIME TO TRY

Let's see how temperature affects the process. Take three clear glass containers of about the same size. Fill one halfway with very hot water, another with water near room temperature, and the third with very cold water. Line them up and wait until the water stops moving. Gently add one drop of food coloring to each container. If you have no food coloring, you could also use a tea bag. Now just observe. You should see evidence that the molecules are diffusing—the color should spread out from where it is most concentrated. Eventually you should also see the equilibrium state—all of the water should be of uniform color, meaning that the molecules have spread out equally, even though they continue to move.

How did the diffusion rates differ between the three containers, and why? ____

____

You should see that increased temperature also increases the diffusion rate. This is because heat makes molecules move faster.

FACILITATED DIFFUSION

Simple diffusion is only one mechanism that allows materials to enter and leave a cell. Only lipid-soluble nonpolar molecules can diffuse directly through the cell membrane. Larger polar molecules, such as glucose (one type of sugar), cannot diffuse through the membrane as easily. Instead, they are moved by a special protein carrier molecule in the cell membrane. The molecules still move from an area of high concentration to one of low concentration, trying to reach equilibrium. The carrier molecule merely helps, or facilitates, the molecules' movements through the cell membrane. It acts like a special door through which they can pass. This type of diffusion is called **facilitated diffusion**.

Diffusion is a passive process by which molecules spontaneously move from where they are in high concentration to where they are in low concentration. ■

OSMOSIS

Osmosis is a special type of diffusion that is also critical to cells' survival. Biologically, it is the diffusion of water through a selectively permeable membrane, such as the cell membrane. We have discussed the lipid nature of the cell membrane, so it seems surprising to discover that water actually passes through it fairly well. This is partly due to the fluid nature of the molecules in the membrane, but for our purposes the specifics are not critical. Before proceeding, though, we do need to recall the basics of solutions.

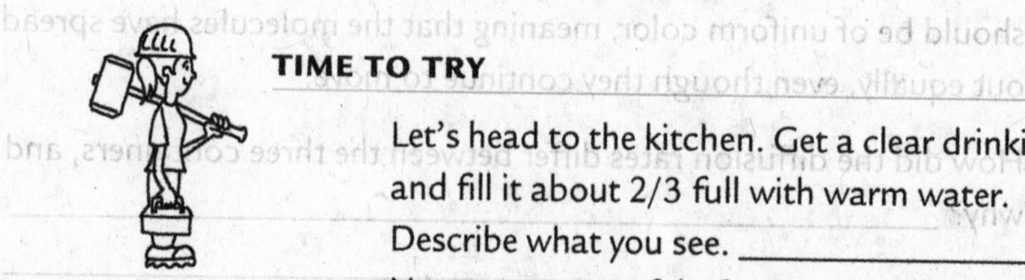

TIME TO TRY

Let's head to the kitchen. Get a clear drinking glass or measuring cup and fill it about 2/3 full with warm water.

Describe what you see. ______________________________

Next, get a spoonful of sugar. Describe what you see. ______________

Finally, add the sugar to the water and mix it thoroughly. What do you see? ______________________________

This exercise may have seemed silly, but you just made a solution by combining a solute and a solvent. The sugar is the **solute**—the material that gets dissolved. As in our bodies, water is the **solvent**—the material that dissolves the solute. The end result is a **solution**, which is a homogeneous mixture—it should look uniformly clear throughout, because you cannot see the dissolved solute.

To understand the relationship between the water and solute molecules, recall that matter is anything that has mass and occupies space, so no two molecules can occupy the same space at the same time. Let's think about that sugar solution you just made. If you started by first putting a half-cup of sugar in the container, there would have been less

FIGURE 6.10 A solution contains both a solvent, such as water, and solute molecules. In this illustration, the beakers contain the same amount of solution. The geometric shapes in each solution represent various solute molecules. The remaining shaded space represents the water. Clearly, beaker A, which contains more solute molecules, has less space left for water, but beaker B, with fewer solute molecules, contains more water.

room left to fill with water. The opposite would also be true—less sugar would mean more water. And it doesn't matter what the solute molecules are—they each take up their own space (**Figure 6.10**). Feel free to use the following highly technical term for the various solutes: *stuff*. When we are talking about osmosis through a cell membrane, we must look at the water and the other stuff for a simple reason—the more stuff you have, the less water there can be because the stuff takes up space that water can no longer occupy.

TIME TO TRY

Prove to yourself that when there are more solute molecules in a given space, there must be less solvent, and vice versa. In other words, the more concentrated the solute, the less concentrated the solvent (water). Get a glass measuring cup, then grab a handful of household items–paper clips, coins, marbles, and so on. The only rules are that the objects cannot float and they cannot dissolve (most food items are not good for this reason). These objects represent the solute molecules—the stuff—and the water, of course, is the solvent. Drop all of the objects into the cup, then fill it with water to the one-cup mark. Look at the layers of water and objects and note the sizes of each layer. Carefully pour the water into a second container and put it aside. Place the objects on some paper towels, then pour the water back into the cup to measure it. How much water was there in the solution? ____________________

Repeat this procedure, but this time place only about half of the items in the cup before filling it with water to the one-cup mark. Again, pour the water into your second container, remove the objects, then pour the water back into the cup to measure it. How much water did your mock "solution" contain this time?

____

In which trial did you use more "solute molecules" (objects)?

____

In which trial did you have the least water? ____________________

Explain the relationship between the amount of solute molecules and the amount of solvent present in a solution. ____________________

____

____

Recall that molecules never stop moving, even after reaching equilibrium. Each of your cells has two solutions separated by a selectively permeable membrane—the extracellular fluid outside of the cell and the intracellular fluid inside the cell. These solutions usually have different compositions. If cell membranes were fully permeable, all molecules could pass through and move down their concentration gradients until reaching equilibrium, and we would have simple diffusion. But the cell membrane is semipermeable. We know water can pass through the cell membrane rather easily, but many solute molecules can't because of their size or their chemical composition. You also know from the previous activity that there is a higher concentration of water (solvent) wherever there is a lower concentration of other stuff (solute molecules).

Look at **Figure 6.11**. In this illustration, there is a higher concentration of solute molecules inside the cell than there is outside. Assume the solute molecules shown cannot pass through the cell membrane. Now answer the following questions:

1. Which fluid contains the lowest concentration of solute molecules?

____

2. Which fluid has the highest concentration of water? (Recall that when water is high, solute is low, and vice versa.)

____

By osmosis, water diffuses—it moves down its concentration gradient—from where it is in higher concentration to where it is in lower concentration. You should see clearly from Figure 6.11 that, in this

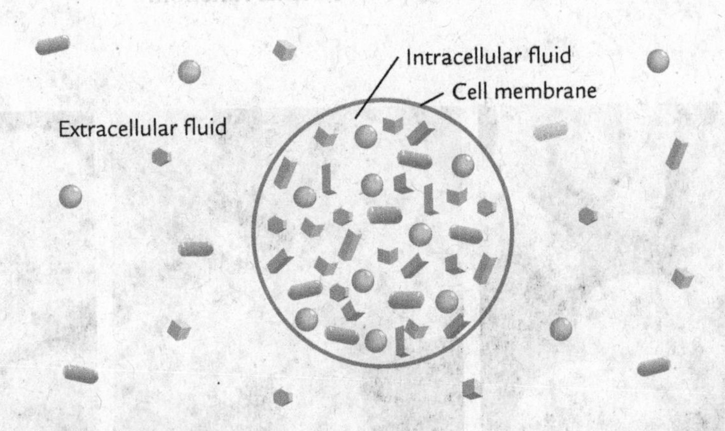

FIGURE 6.11 The cell membrane separates the extracellular fluid from the intracellular fluid. In this figure, a situation is shown in which there are more solute molecules inside the cell than there are outside of it. Assume the shaded shapes are solute molecules and the remaining space is filled with water.

example, there is a greater concentration of water in the extracellular fluid than there is in the intracellular fluid, so water will enter the cell.

For osmosis, you just need to know where the highest concentration of water is, because it will always move from that area to the area with less concentration. The only reason to consider the other stuff (solute) is because it tells you where the water is—the area with the lower solute concentration has the higher water concentration, so the water will move away from that area.

Osmosis is diffusion of water through a selectively permeable membrane. ■

What happens to cells when water moves across the cell membrane? **Figure 6.12** shows three views of red blood cells. Figure 6.12a shows a normal red blood cell, shaped like a biconcave disk. If water leaves a cell, the cell will shrink and the cell won't function efficiently. This is shown in Figure 6.12b. If water moves into a cell, the cell will swell. This is shown in Figure 6.12c. This swelling increases the pressure inside the cell and impedes normal function.

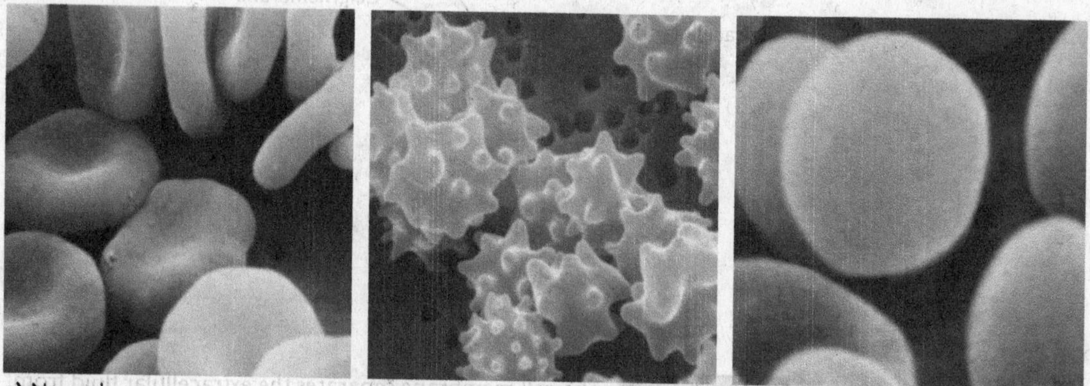

a) Normal **b)** Too little water **c)** Too much water

FIGURE 6.12 Red blood cells showing the effects of osmosis. **a)** Normal red blood cells. **b)** Red blood cells that have lost water through osmosis. **c)** Red blood cells swollen from taking in too much water by osmosis, and at risk of rupturing.

PICTURE THIS

You are already familiar with water moving into and out of cells and changing their shapes and sizes. As you read earlier, plant cells are surrounded by a rigid cell wall that helps maintain their shape. Have you ever had a houseplant or a plant in your yard that was a bit wilted and droopy? When the plant cells lose water, they shrink and cannot support the weight of the plant parts above them, so they droop. When they receive water, the cells swell, exerting pressure against the rigid cell wall, and this pressure helps support the rest of the plant—it stands back up.

WHY SHOULD I CARE?

Animal cells lack a cell wall, so they are more likely than plant cells to rupture if they take in too much water. In health care, IV (intravenous) fluids must have the appropriate amount of water—if there is more water in the IV fluid than in the patient's intracellular fluid, his or her cells will swell. If there is less water in the IV fluid, the patient's cells will shrink. Unless the patient is being treated for a fluid imbalance, IV fluids should have the same water concentration as intracellular fluids so that the water will be at equilibrium. Such fluids are called *isotonic*. The wrong amount of water in IV fluid can be fatal.

✔ QUICK CHECK

Refer back to Figure 6.11. Will water enter or leave the cell?

Answer: The water moves from where it is more concentrated to where it is less concentrated, so it moves *into* the cell.

ACTIVE TRANSPORT

It took awhile to get through osmosis, so let's regroup and review what we have done. We discussed simple diffusion, facilitated diffusion, and osmosis. These are all types of diffusion, so the molecules being examined will move from an area where they are more concentrated to an area where they are less concentrated. Diffusion is a spontaneous process—it happens automatically. These types of diffusion do not require energy, so they are referred to as **passive transport.**

Sometimes, instead, cells need to move molecules against their concentration gradients. For example, for a nerve cell to transmit a signal, sodium ions must enter the cell. Special channels in the cell membrane open and sodium ions diffuse into the cell because there are more sodium ions outside than inside. This is simple diffusion. After that nerve signal is sent, though, the sodium ions must leave so the cell is ready for the next signal. But there are still more sodium ions outside than there are inside. The ones inside can't merely diffuse out—they would be moving up their concentration gradient.

PICTURE THIS

Assume you are a dedicated biology student and you've been studying much more than you've been cleaning house. You hear a van pull up outside and you see that it is the *Prize Patrol* from a major sweepstakes sponsor, with a camera crew. They are broadcasting live and they are quickly approaching your door. You grab an armfull of clutter and all your books and open your closet door to stash them on the overhead shelf, but the shelf is full because it has already been jammed with previous clutter.

1. What likely happens if you just open the closet door and stand back? ____________________

2. Is it easy or difficult to cram more clutter onto the closet shelf?

   ____

3. Does the clutter spontaneously head onto the shelf on its own, or do you have to really work at getting it to go and stay there?

   ____

In the situation with the closet, when you open the door there is more clutter inside on the shelf than there is outside, so the clutter inside will spontaneously fall out. This represents diffusion—molecules moving from where they are more concentrated to where they are less concentrated. You likely have to use your arm or in some way exert effort to prevent the clutter from falling out. At the same time, you have to force the additional clutter in your arms into the closet, and you have to work hard to do so. Work requires energy, so you must use energy to accomplish the task.

With cells, molecules will not spontaneously move against their concentration gradients. Work must be done, so energy must be used. For this reason, this type of movement—moving molecules up, or against, their concentration gradient—is called **active transport**. This type of movement also requires special one-way "doorways" in the cell membrane, called pumps, that ensure the molecules can only move in one direction. Otherwise molecules on the opposite side would spontaneously diffuse out, which is not our goal.

Active transport moves molecules against their concentration gradient, which requires a special molecular pump and energy. ■

EXOCYTOSIS

We have discussed ways for moving molecules, but now it is time to think bigger! Recall that cells may make proteins and other molecules that will be exported. These products are usually wrapped in a membranous sac, called a vesicle, at the Golgi apparatus. The method by which they are expelled is called **exocytosis** (*exo-* = outside, *cyto-* = cell). The method is rather simple. The vesicle makes its way to the edge of the cell and its membrane fuses with the cell membrane. As the vesicle pushes against the cell membrane, its own membrane ruptures and seems to peel back, becoming part of the cell membrane and releasing its contents. **Figure 6.13** shows a cell secreting a product via exocytosis. Notice the contents spewing out of the cell.

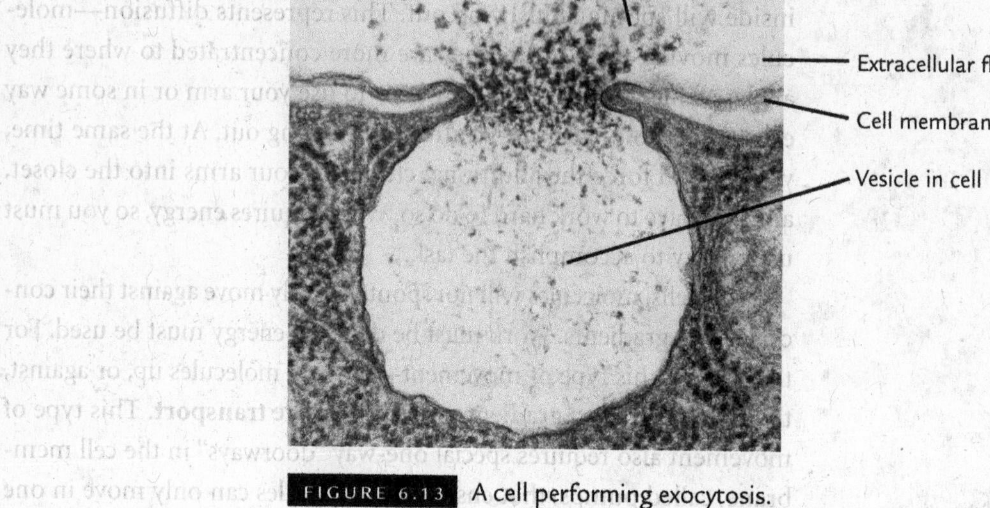

FIGURE 6.13 A cell performing exocytosis.

ENDOCYTOSIS

Endocytosis (*endo-* = inside) is the reverse of exocytosis. It is a means by which cells can take in rather large objects or even liquid that contains dissolved materials, such as nutrients. There are three major types of endocytosis:

- phagocytosis,
- receptor-mediated endocytosis, and
- pinocytosis.

Phagocytosis is the process by which solids are moved into your cells, and is sometimes referred to as "cell eating." This process demonstrates the active nature of the cell membrane. Extensions of the cell membrane, called **pseudopodia** ("false feet"), seem to reach out from the cell surface, rather like tiny arms, on each side of the object to be taken in. Then the extensions fuse and form a membranous sac, like a vesicle, around the object. This saclike structure is called a **phagosome**, and it moves inward, pinching off from the cell membrane on the inside of the cell. The object is now inside the cell. Soon, several lysosomes typically fuse with the phagosome. Some cells in your immune system are specialized for phagocytosis and use this process to rid your body of foreign material, such as bacteria, that might make you ill.

✔ **QUICK CHECK**

Why do lysosomes fuse with the phagosome? ______________________

____

Answer: Lysosomes contain enzymes that will break down the contents of the phagosome, rendering potential threats harmless and recycling materials that are then made available to the cell for reuse.

Receptor-mediated endocytosis is related to phagocytosis, but before the cell "reaches out" to take the materials in, the objects to be moved must bind to special receptors in the cell membrane. This binding triggers the cell membrane to form a pocket that will enclose the materials and bring them into the cell. From this point on, the process is very much like phagocytosis.

Pinocytosis is a way cells can bring in liquids, and is sometimes referred to as "cell drinking." In this process, part of the cell membrane puckers inward, forming a pouch at the surface, and the pouch contains extracellular fluid. In that fluid are a variety of dissolved substances that are now surrounded by cell membrane, which pinches off to form a vesicle-like structure called an **endosome**. The contents are then released inside the cell and are available for its use.

Ashes to Ashes, Cells to Cells: **The Cell Cycle**

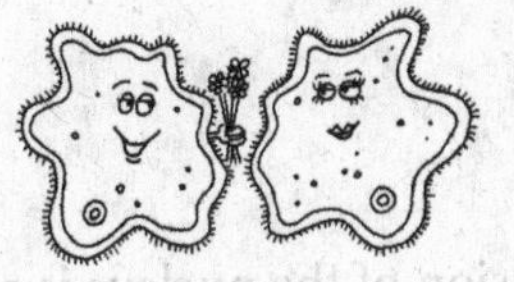

We have discussed many processes that occur in a cell, but a very important one remains: the cell life cycle. Remember from the cell theory that cells come from preexisting cells. Each cell goes through the **cell cycle** (**Figure 6.14**), and the duration of this cycle varies with the cell type. The cell cycle is divided into multiple phases. Simply, it can be viewed as having two main parts: **interphase** and **cell reproduction.**

Most cells spend the majority of their life cycle in interphase, which is when they are doing their normal living activities. This makes sense—

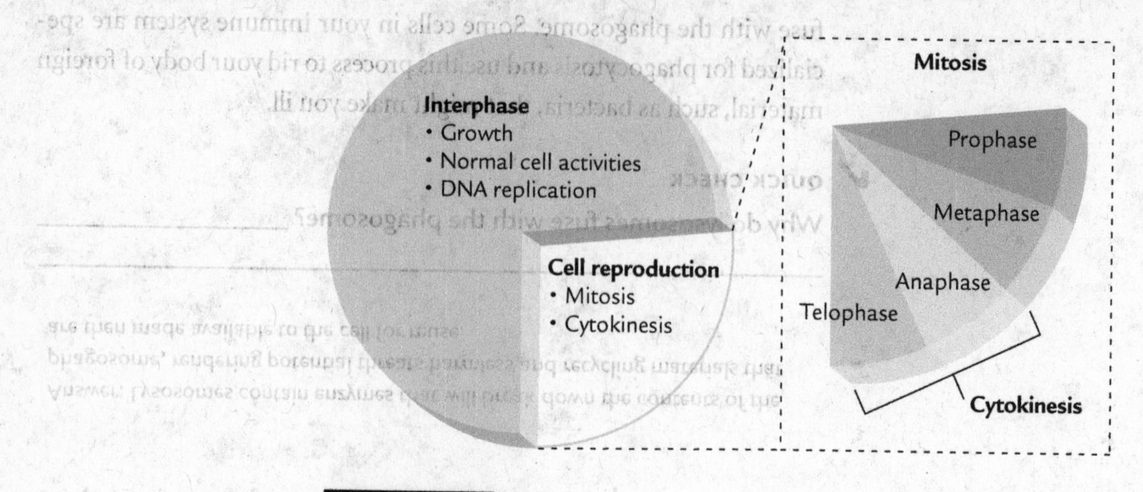

FIGURE 6.14 The cell cycle.

we spend most of our time living our life, and only a fraction, if any, in reproductive behavior. During interphase, cells go about their normal business, growing, maturing, and doing all the activities we have discussed and more. This is when the cell makes its contribution to the overall function of the whole organism.

During this phase, it also prepares for the reproduction to come. The DNA—the genetic material—replicates. That means it reproduces. Cells reproduce by dividing, but each daughter cell needs to have a complete set of all the DNA. So, during interphase, the complete set of DNA is copied in a process called **DNA replication**.

Once the DNA has replicated, cell reproduction may begin. Cell division, or reproduction, includes two processes:

- **mitosis**, which is nuclear division, and
- **cytokinesis**, which is cytoplasmic division.

The DNA is important to your cells, and division of the nucleus is a separate process, called **mitosis**. This is a very precise event designed to ensure that the DNA is equally divided into the daughter cells. Mitosis includes four phases:

- prophase,
- metaphase,
- anaphase, and
- telophase.

During interphase, the DNA is stretched out in thin strands called chromatin. During **prophase**, the first phase of mitosis, the chromatin condenses into rodlike structures—the chromosomes. The nuclear membrane also disappears so that the chromosomes can move more freely. The next phase is **metaphase**. *Meta-* means *middle*, and during metaphase the chromosomes align very precisely, in duplicated pairs, along the midline, or equator, of the cell. This ensures that they will separate precisely. The next phase is **anaphase**. *Ana-* means *away*, and the duplicated chromosomes separate during this phase—one complete set goes to each side, or pole, of the cell. Movement of the chromosomes is directed by the centrioles and delicate structures that form from them called spindle fibers, which pull the chromosomes in opposite directions. The final phase of mitosis is **telophase**, during which the chromosomes complete their journey to the opposite poles. This phase is like a reverse prophase. The nuclear membrane reappears and the chromosomes relax back into the stretched-out chromatin strands. See **Table 6.2** to learn more about the stages of the cell cycle.

TIME TO TRY

In the spaces provided, list the phases of mitosis in the correct order, then sketch what the cell would look like during that phase.

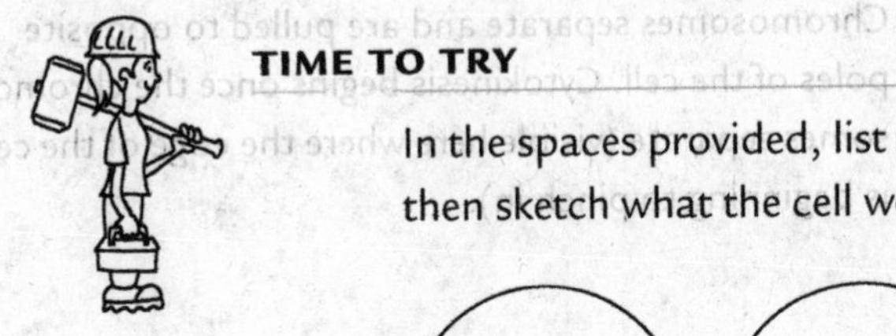

______ ______ ______ ______

TABLE 6.2 The stages of the cell cycle include interphase and cell reproduction. Cell reproduction includes the four phases of mitosis during which the nuclear contents divide, and cytokinesis during which the remainder of the cell divides. Cytokinesis overlaps the latter phases of mitosis.

Picture	Stage	Events
	Interphase	Normal cell activities, growth, DNA replication. DNA is visible as thin strands called chromatin.
	Prophase	Chromatin condenses into rodlike structures called chromosomes that are clearly visible, and the nuclear membrane disappears.
	Metaphase	Chromosomes align very precisely along the midline.
	Anaphase	Chromosomes separate and are pulled to opposite poles of the cell. Cytokinesis begins once the chromosomes separate (visible here where the edge of the cell is beginning to pinch in).
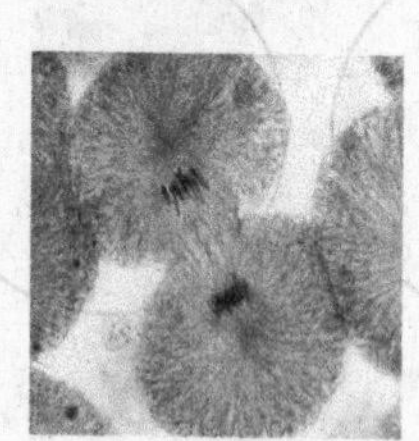	Telophase	Chromosomes are in opposite poles and cytokinesis continues. This stage ends when the cells completely separate, forming two daughter cells.

Mitosis just divides the nucleus or, more specifically, the chromosomes. Once the chromosomes have carefully separated, the rest of the cell can be divided. This process, by which the cytoplasm divides, is called **cytokinesis**. It begins during anaphase of mitosis, after the chromosomes have separated, and ends at the end of telophase. At the end of cytokinesis, the original cell is gone and in its wake are two new daughter cells, almost identical and each with a complete set of DNA. These daughter cells are in interphase, and the whole cell cycle begins anew.

Cell reproduction includes mitosis, which is division of the nuclear contents, and cytokinesis, which is division of the cytoplasm. ■

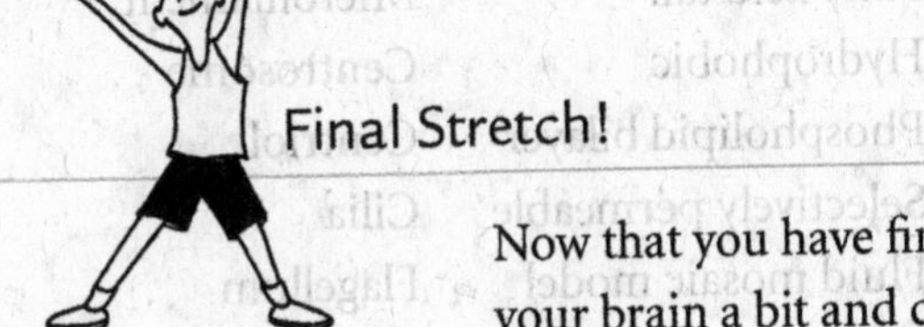

Final Stretch!

Now that you have finished reading this chapter, it is time to stretch your brain a bit and check how much you learned.

RUNNING WORDS

At the end of each chapter, be sure you have learned the language. Here are the terms introduced in this chapter with which you should be familiar. Write them in a notebook and define them in your own words, then go back through the chapter to check your meaning, correcting as needed. Also try to list examples when appropriate.

Cell
Cell theory
Unicellular
Multicellular
Cell membrane
Cytoplasm
DNA
Nucleoid
Plasmid
Homeostasis
Plasma membrane
Phospholipid
Phosphate head
Hydrophilic
Fatty acid tail
Hydrophobic
Phospholipid bilayer
Selectively permeable
Fluid mosaic model
Cytosol
Nucleus
Chromatin
Chromosome
Nuclear envelope
Nuclear pore
Nucleoplasm
Ribosome
Protein synthesis
Messenger RNA (mRNA)
Ribosomal RNA (rRNA)
Nucleolus
Transfer RNA (tRNA)
Endoplasmic reticulum (ER)
Golgi apparatus
Vesicle
Mitochondrion
ATP
Chloroplast
Lysosome
Cytoskeleton
Microtubule
Microfilament
Centrosome
Centriole
Cilia
Flagellum
Brownian motion
Concentration gradient
Equilibrium
Simple diffusion
Facilitated diffusion
Osmosis
Solute
Solvent
Solution
Passive transport

Active transport	Phagosome	Cell cycle	Prophase
Exocytosis	Receptor-mediated endocytosis	Interphase	Metaphase
Endocytosis	Pinocytosis	Cell reproduction	Anaphase
Phagocytosis	Endosome	DNA replication	Telophase
Pseudopodia		Mitosis	Cytokinesis

WHAT DID YOU LEARN?

PART A: ANSWER THE FOLLOWING QUESTIONS.

1. List the five principles of the cell theory.

____

____

____

____

____

2. What is the basic difference between prokaryotic cells and eukaryotic cells?

____

____

3. Describe the organization of the cell membrane. _______________

____

____

4. Design a concept map for the following terms: protein, DNA, nucleus, nucleolus, ribosome, rough ER, Golgi apparatus, and exocytosis.

5. Differentiate between passive and active movement processes. ____________________

6. In terms of movement processes, explain how making a cup of hot tea with a tea bag involves both osmosis and simple diffusion. ____________________

7. Assume you have limp carrot sticks in your refrigerator. To get them plump and crisp again, would you soak them in pure water or in a strong salt solution?

8. Differentiate between phagocytosis and pinoctyosis. ____________________

9. List in order the four phases of mitosis: ____________________

10. Explain the complete cell cycle. ____________________

PART B: FOR EACH OF THE FOLLOWING ITEMS, MATCH THE TERM WITH ITS DESCRIPTION.

1. mitochondrion ________
2. lysosome ________
3. ribosome ________
4. Golgi apparatus ________
5. nucleolus ________
6. chloroplast ________
7. exocytosis ________
8. osmosis ________
9. active transport ________
10. simple diffusion ________

a) Diffusion of water through a selectively permeable membrane.
b) Expelling materials out of the cell.
c) Site where ribosomes are made.
d) Process that uses energy to move molecules against their concentration gradients.
e) Spontaneous movement of molecules down their concentration gradient.
f) "Garbage disposal" containing digestive enzymes.
g) Site where proteins are made.
h) Site where cellular products are packaged.
i) Site of photosynthesis.
j) Site where ATP (energy) is made.

CHAPTER 1

Answers will vary from student to student.

CHAPTER 2

Part A

1. –32
2. 20,000
3. 3/4
4. 1/4
5. 1/4
6. 0.2
7. 720 breaths per hour
8. 400 cm
9. 30.48 cm
10. 120 lbs

Part B

1. 14
2. 4
3. 0.3; 30%
4. 20%
5. volume of a liquid
6. gram; meter (technically kilometer, but in practical use it is the meter); liter
7. "Normal" blood pressure is the average blood pressure.
8. 100°C
9. volume
10. 1000

CHAPTER 3

Part A

1. e
2. f
3. i
4. j
5. g
6. h
7. b
8. d
9. c
10. a

Part B

1. d
2. b
3. glycolysis; exocytosis; epiphyte
4. The process of dividing the cytoplasm of a cell; "colored body"—condensed genetic material that can be stained and seen under a microscope; plant growth in response to touch.
5. pharynges; mitochondria; coxae

CHAPTER 4

Part A

1. ecosystem, community, population, organism, organ system, organ, tissue, cell, organelle, macromolecule, molecule, atom
2. The foot is a broadened surface with arches that distribute the weight evenly to all parts of the foot that contact the ground.
3. Homeostasis is the maintenance of a constant internal environment. It is important to maintain homeostasis because cells perform optimally in a balanced internal environment.
4. DNA transmits information about characteristics from parents to offspring.
5. Most living things ultimately get their energy from the sun.
6. When a population of insects encounters pesticides in its environment, some insects will survive the exposure and live to produce offspring. These offspring will have the pesticide-resistant characteristics of their parents. Over time, the population of insects will contain a larger percentage of the pesticide-resistant individuals. This is an example of natural selection in action.
7. Domain Bacteria, domain Archaea, domain Eukarya; kingdom Protista, kingdom Plantae, kingdom Fungi, kingdom Animalia.
8. Inductive reasoning is done from a set of specific observations to a general conclusion, whereas deductive reasoning is done by extrapolating specifics from a general premise.

Part B

1. c
2. g
3. f
4. a
5. e
6. b
7. d
8. h

CHAPTER 5

Part A

	Potassium	Iodine	Oxygen	Neon
Chemical symbol	K	I	O	Ne
Atomic number	19	53	8	10
Atomic weight	39.10	126.90	16.00	20.18
Number of protons	19	53	8	10
Number of electrons	19	53	8	10
Number of neutrons	20	74	8	10
Number of electrons in outermost shell	1	7	6	8

Part B

1. protons and neutrons
2. electrons
3. protons
4. carbon, hydrogen, oxygen, nitrogen
5. The row tells you how many shells of electrons that element has.
6. The column tells you how many electrons are in the outer shell of that element.
7. The elements in the far right column are stable because they have a full outermost shell of electrons.
8. 2; 8
9. The calcium atom lost 2 electrons, giving it a $^{+}2$ charge.
10. In ionic bonds, the electrons physically move from one atom to another so that the atoms involved lose or gain electrons, and the opposite charges of the resulting ions draw the atoms together. In covalent bonding, the atoms share electrons.
11. structural
12. decomposition (catabolic)
13. synthetic (anabolic)
14. carbon and hydrogen
15. $C_6H_{12}O_6$ <u>O</u>
    CO_2 <u>I</u>
    CH_4 <u>O</u>
    CO <u>I</u>
    HCl <u>I</u>
    H_2O <u>I</u>

CHAPTER 6

Part A

1. All organisms are composed of one or more cells; cells are the basic structural and functional units of life; all vital functions of an organism occur within cells; all cells come from preexisting cells; and cells contain hereditary information that regulates cell functions and is passed from generation to generation.
2. Prokaryotes have no nucleus and lack internal membranes.
3. The cell membrane is a phospholipid bilayer with protein channels periodically passing through it and other molecules such as cholesterol and carbohydrates "floating" in it.
4. Concept maps will vary from student to student, but these relationships should be included:
   - DNA contains the instructions for how to build proteins and is located in the nucleus.
   - The nucleolus is located inside the nucleus and makes ribosomes.
   - Rough endoplasmic reticulum gets its appearance from the presence of ribosomes.
   - Proteins are made at ribosomes, so rough ER makes proteins.
   - Proteins from rough ER move to the Golgi apparatus for processing and packaging, then are expelled from the cell by exocytosis.
5. In passive movement processes, substances move along their concentration gradients from high concentration to low concentration. In active movement processes (active transport), energy is used to move substances against their concentration gradients (from low concentration to high).

6. When a tea bag is placed in hot water, the water moves into the tea bag by osmosis and dissolves the tea, then the tea molecules diffuse out of the tea leaves into the water.
7. Water would move into the carrot, plumping the cells and making them rigid. (Salt would draw more water out, causing them to shrivel even more.)
8. In phagocytosis, the cell surrounds a solid, with the cell membrane extending out to surround the object, then drawing it inward. In pinocytosis, the cell membrane forms an inward pouch that surrounds a droplet of liquid.
9. prophase, metaphase, anaphase, and telophase
10. The complete cell cycle includes all events that occur in the life of a cell, specifically interphase and cell reproduction. Interphase is when the cell carries out its normal functions, and the DNA replicates in preparation for reproduction. Cell reproduction includes the four phases of mitosis that divide the nucleus, and cytokinesis that divides the cytoplasm.

Part B

1. j
2. f
3. g
4. h
5. c
6. i
7. b
8. a
9. d
10. e

Index

Photo and Illustration Credits

Unless noted, all chapter opener art and cartoon spot art was created by Kevin Opstedal.

CHAPTER 1

Table 1.2: All photos from PhotoDisc.

Figures 1.1, 1.2, 1.3, 1.4: Seventeenth Street Studios.

Figure 1.5: Tom Steward/CORBIS.

CHAPTER 2

Figures 2.1, 2.2, 2.3, 2.4, 2.5: Seventeenth Street Studios.

Figure 2.6: Benjamin Cummings Publishing, Pearson Education.

Figure 2.7a: Ohaus Corporation.

Figure 2.7b: Seventeenth Street Studios.

Figure 2.8: Richard Megna/ Fundamental Photographs.

Figures 2.9, 2.10, 2.11: Seventeenth Street Studios.

CHAPTER 3

Figures 3.1, 3.2: Seventeenth Street Studios.

Figure 3.3; hip implant: National Institutes of Health. Art by Seventeenth Street Studios.

Figure 3.4: Victoria & Albert Museum, London/Art Resource, NY.

CHAPTER 4

Figure 4.1: PhotoDisc/Getty Images; James Gritz; PhotoDisc/Getty Images/Adapted from Campbell, Reece, and Simon, *Essential Biology,* 3e, F1.3, © Benjamin Cummings, 2007.

Figure 4.2: Seventeenth Street Studios.

Figure 4.3a: VideoSurgery/Science Source.

Figure 4.3b: VideoSurgery/Science Source.

Figure 4.4: Adapted from Campbell, Reece, and Simon, *Essential Biology,* 3e, F1.4, © Benjamin Cummings, 2007.

Figure 4.5a: Premaphotos/Alamy.

Figure 4.5b: John Cancalosi/National Geographic Society/Corbis/ Corbis.

Figure 4.5c: Adegsm/Moment/Getty Images.

Figure 4.6: Adapted from Campbell and Reece, *Biology,* 7e, F1.21, © Benjamin Cummings, 2004.

Figure 4.7a: Oliver Meckes/Nicole Ottawa/Photo Researchers/ Adapted from Campbell, Reece, and Simon, *Essential Biology,* 3e, F1.9, © Benjamin Cummings, 2007.

Figure 4.7b: Eye of Science/Photo Researchers, Inc.

Figure 4.7c: CORBIS/Adapted from Campbell, Reece, and Simon, *Essential Biology,* 3e, F1.9, © Benjamin Cummings, 2007.

Figure 4.7d: Digital Vision/Getty Images/Adapted from Campbell, Reece, and Simon, *Essential Biology,* 3e, F1.9, © Benjamin Cummings, 2007.

Figure 4.7e: CORBIS/Adapted from Campbell, Reece, and Simon, *Essential Biology*, 3e, F1.9, © Benjamin Cummings, 2007.

Figure 4.7f: D.P. Wilson/Photo Researchers/Adapted from Campbell, Reece, and Simon, *Essential Biology*, 3e, F1.9, © Benjamin Cummings, 2007.

Figure 4.8: Adapted from Campbell, Reece, and Simon, *Essential Biology*, 3e, F1.19, © Benjamin Cummings, 2007.

CHAPTER 5

Figure 5.1: PhotoDisc/Getty Images; James Gritz; PhotoDisc/Getty Images/Adapted from Campbell, Reece, and Simon, *Essential Biology,* 3e, F1.3, © Benjamin Cummings, 2007.

Figures 5.2, 5.3, 5.4, Time to Try, 5.5, 5.6, 5.7, 5.8, 5.9, 5.10: Seventeenth Street Studios.

Figure 5.8: Lori Garrett.

CHAPTER 6

Figure 6.1: PhotoDisc/Getty Images; James Gritz; PhotoDisc/Getty Images/Adapted from Campbell, Reece, and Simon, *Essential Biology,* 3e, F1.3, © Benjamin Cummings, 2007.

Figure 6.2a: Cheryl Power/Photo Researchers, Inc.

Figure 6.2b: David McCarthy/ Photo Researchers.

Figure 6.2c: David M. Phillips/ Photo Researchers.

Figure 6.3, 6.4: Seventeenth Street Studios.

Figure 6.5: Adapted from Marieb, *Study Guide for Human Anatomy and Physiology,* 6e, F3.1, © Benjamin Cummings, 2004.

Figure 6.6: Adapted from Campbell, Reece, and Sim on, *Essential Biology,* 3e, F4.6, © Benjamin Cummings, 2007.

Figure 6.7: D.W. Fawcett/Photo Researchers/Adapted from Bauman, *Microbiology*, F3.37, © Benjamin Cummings, 2004.

Figure 6.8: Adapted from Campbell, Reece, and Simon, *Essential Biology*, 3e, F4.16, © Benjamin Cummings, 2007.

Figure 6.9, 6.10, 6.11: Seventeenth Street Studios.

Figure 6.12a, b, c: M. Sheetz, R. Painter, and S. Singer, *Journal of Cell Biology* 1976. 70: 193.

Figure 6.13: Birgit H. Satir, Dept. of Anatomy and Structural Biology, Albert Einstein College of Medicine.

Figure 6.14: Seventeenth Street Studios.

Table 6.2: All photos by Ed Reschke.

Notes

Notes

Notes